ROUTLEDGE LIBRARY EDITIONS:
POLITICAL GEOGRAPHY

Volume 9

T0265111

THE GEOGRAPHY OF WARFARE

THE GEOGRAPHY OF WARFARE

PATRICK O'SULLIVAN AND
JESSE W. MILLER JR

LONDON AND NEW YORK

First published in 1983

This edition first published in 2015
by Routledge
2 Park Square, Milton Park, Abingdon, Oxon, OX14 4RN

and by Routledge
711 Third Avenue, New York, NY 10017

Routledge is an imprint of the Taylor & Francis Group, an informa business

© 1983 Patrick O'Sullivan and Jesse W. Miller Jr

British Library Cataloguing in Publication Data
A catalogue record for this book is available from the British Library

ISBN: 978-1-138-80830-0 (Set)
eISBN: 978-1-315-74725-5 (Set)
ISBN: 978-1-138-81057-0 (Volume 9)
eISBN: 978-1-315-74942-6 (Volume 9)
Pb ISBN: 978-1-138-81058-7 (Volume 9)

Publisher's Note
The publisher has gone to great lengths to ensure the quality of this reprint but points out that some imperfections in the original copies may be apparent.

Disclaimer
The publisher has made every effort to trace copyright holders and would welcome correspondence from those they have been unable to trace.

Printed and bound by CPI Group (UK) Ltd, Croydon, CR0 4YY

The GEOGRAPHY OF WARFARE

PATRICK O'SULLIVAN AND JESSE W. MILLER JR

CROOM HELM
London & Canberra

© 1983 Patrick O'Sullivan and Jesse W. Miller Jr
Croom Helm Ltd, Provident House, Burrell Row, Beckenham, Kent BR3 1AT

British Library Cataloguing in Publication Data

O'Sullivan, Patrick
    The geography of warfare.
    1. Military art and science—History
    2. Naval art and science—History
    I. Title   II. Miller, Jesse N.
    355'009   U27

ISBN 0-7099-1918-2

Printed and bound in Great Britain
by Billing and Sons Ltd, Worcester.

CONTENTS

1 GEOGRAPHY, TACTICS AND STRATEGY

Therefore to estimate the enemy situation and to calculate the distances and degree of difficulty of the terrain so as to control victory are virtues of the superior general

Sun Tzu, (chapter 10, verse 17).

Introduction

The justification for writing this book is that the fundamental strategic and tactical problems are geographical in nature. Field Marshal Montgomery once attributed victory in battle to 'transportation, administration and geography', with the accent on the latter. It is not our purpose to describe the geographical disposition of armed force or to record the applications of geography to the conduct of military affairs. This has been done admirably elsewhere. Our theme is the geography of preparing for and waging war. The decision whether to fight or not ought to be informed by a keen sense of geopolitical realities. Although the question of how to fight is governed by technological and economic capabilities, it is essentially a response to environmental possibilities and limitations. Once if and how to fight have been determined, the problems of war become much more specifically geographical and the principal matter for decision is where to commit forces to battle.

We can distinguish three activities involved in the prosecution of war. Firstly, information has to be gathered. Where are the objectives to be captured or defended? Where is the opposition likely to come from or be? Where are there obstacles and channels of movement? Secondly, the commitment of force can only be achieved within the domain of feasibility of logistics. The possibilities for action are limited by where supplies of men, material and firepower can be deployed. Thirdly, after intelligence and logistics have provided information on the geography of the problem and the logistically favourable ambit then decisions on action are a matter of where to commit what force? where to move and deploy? where to attack or defend, advance or retreat? where to place routes, bridges, landings and defences?

The relations between intelligence, logistics and action can be viewed in terms of geographically specific informational and logistic constraints on the freedom of action. These relations operate at a variety of geogra-

phical scales with what is feasible at a more local scope limiting freedom of choice at more global levels. We can distinguish four categories
of scale related to kinds of decision and action:

(a) local – concerned with operational, tactical decisions below the
division level

(b) operational – the manoeuvre and support of major field forces

(c) regional – concerned with strategic deployment in a theatre of
war, campaign considerations

(d) global – matters of grand strategy, concerned with geopolitical
goals and ploys and the design of worldwide strategic systems.

The Scope and Scale of Warfare

There is really no ready cut-off between the very local considerations
of what troops can do on the ground with their weapons and the use of
terrain and the broader prospect of the deployment of troops and
weapons at the global scale. Tactical and strategic questions merge into
each other. Tactics and strategy are usually defined in terms of each
other, like boats and ships. Strategy merely implies a greater level of
generality than tactics. Conventionally, strategy has been taken to
imply generalship, the art of conducting a campaign and manoeuvering
an army conceived at a fairly high level of geographical abstraction. The
image is of broad arrows moving large units through a greatly simplified
landscape. Above this we have grand strategy with a broader, if not
worldwide, connotation. This implies the use not only of military but
also economic, diplomatic and political means to achieve national ends.
By contrast, tactics is the art of manoeuvring and coordinating the fires
of land, sea or air forces in the presence of the enemy and implies a
detailed scrutiny of local geography and specifying the deployment of
small individual units.

Clearly, geographical factors come into play at all levels. Global
strategies must be informed by geographic intelligence and the highest
level decisions are fundamentally geopolitical in nature and can go
awry from geographical misconception. If arms are to be employed in
seeking a geopolitical goal, then the advantages of the strategic situation
are clearly conditioned by what is feasible at the tactical level. The
technology or mobility of the deployment of firepower and communications gives the advantage of the various theatres of war to aggression

or defence. The balance of vulnerability and penetrating force can be changed by changes in weapons, mobility, communications or, indeed, terrain. Obviously changes in the balance of aggressive or defensive advantage may change the balance of military power and, thus, the terms of grand strategy and global dominance. It is evident that these scales and levels of decision blend into one another. Nevertheless, there are distinctly different problems at the different geographical scales and there is an hierarchical nesting of their relevance. Campaign strategy is governed by the grand strategic design and in its turn prescribes operations and tactics. What can be done at the global scale is limited by what can be achieved in particular theatres of war, which depends upon the successful execution of tactics. With this in mind, it seems appropriate to build our discussion of the geography of warfare from the ground upwards with the local, operational and tactical as a basis for broader scopes of concern. To set the scene for the relationship of tactics to terrain we need to discuss matters of intelligence, logistics and the variety of tactical ploys. The core of the book then is a treatment of tactics and terrain in conventional warfare under a variety of environmental conditions. From this we step up to the strategic level of resolution and the management of war over a broad theatre. There is a need to concern ourselves with kinds of strategy, the informational needs and logistical considerations involved in a war rather than a battle. We distinguish this geographic scale as that of campaign strategy. It is at this scale that sea, air and nuclear warfare need particular attention.

The treatment of seapower and airpower leads us into the issues of geopolitics and grand strategy. The choices and actions available to pursue global aims are conditioned by the limits to manoeuvre and operational capability, and this ties back into our consideration of the design of global logistic and weapon systems. This discussion sets the scene for the expanded examination of two increasingly important unconventional varieties of fighting — guerrilla warfare and urban warfare. The final chapter brings all of what has gone before to bear in identifying the potential flashpoints and zones of conflict of the earth and attends to the problems of uncertainty and prediction in these matters.

Before setting out on this investigation of the geography of warfare, it is necessary to set the whole matter in its broadest historical setting, considering not merely the evolution of weapons and military organisation, but also its geographic relationship to political and economic structure and the locus of power.

Military Technology and the Geography of Power

Although our civilisation was founded in peaceful cooperation it provided the temptation for political violence. The centralised control over
an agricultural surplus signalled by the appearance in the landscape of
the cities of Sumeria was soon followed by an investment of energy in
war, metal helmets, spearheads and shields to do battle with each other
and with barbarian raiders. By 3000 BC inter-city warfare was already
frequent and semisecular 'kingship' was borrowed from Semitic
pastoralists, encroaching on the essentially pacific temple communities
which had organised these societies from their inception. Concentration
of power in the hands of a single man did something to improve local
defence against barbarian raids, building massive city walls, but
struggles among the cities themselves became sharper and absorbed an
increasing proportion of social energy. 'Power ingests weaker centers
of power or stimulates rival centers to strengthen themselves' (McNeill,
1963). The assimilation of Sumerian technology by the Semitic
Akkadians with their warlike pastoral ways, meant that irrigation technology now spread rapidly after 2000 BC as military force was used to
organise labour for irrigation works. The dilemma presented by the task
of supplying a large professional army from an extensive agricultural
base, while maintaining a central control over the troops, was solved by
1700 BC. Hammurabi of Babylon had a bureaucracy which recorded
the names, locations and duties of thousands of soldiers scattered on
land holdings throughout his realm.

By contrast, Egypt, the first offspring of Mesopotamian civilisation
was insulated from barbarian incursions by desert and had in the Nile
an artery that flowed calmly north with prevailing winds allowing of
sailing against the flow. By controlling shipping a ruler could effectively regulate the entire extent of surplus cultivation. Rather than
having to elaborate an administrative organisation the river provided
the articulation for the early and lasting formation of a unitary state.
In Egypt trade was managed by the royal household and was a military function. Expeditions were mounted to get timber from Syria,
copper from Sinai, gold from Nubia and taxes from the countryside.
From the beginning of civilisation then, there has been an intimate
relationship between politics, the military and commerce.

A major stimulus to increased social and political complexity came with the fusion of the culture of these urban civilisations with the techniques and attitudes of the warrior herdsmen of the western steppes of Eurasia. The military expression of this fusion was the invention, around 1700 BC, of the light, two-wheeled war chariot — soon to become the vehicle of victory throughout Eurasia. The steppe people domesticated horses around 3000 BC. The hitching of horses to wheeled vehicles stemmed from Mesopotamia with Sumerian origins. Their four-wheeled carts could only corner by dragging around and were unsuited to war. The speed and ease of turning needed for fighting was achieved by designing a two-wheeled rig with light, spoked wheels and harness which spread part of the weight of the vehicle to the horses. The primary component of this new style armament was this chariot, but along with it went a powerful compound bow and the practice of making rectangular earthwork fortifications in open country. Mobile firepower and rapid pursuit threw bronze-armoured, massed foot soldiers into disarray. To forestall surprise attack of an encampment of charioteers, a quadrilateral of earth was thrown up (thereby providing town planning with its basic model).

As they adopted the new mode of fighting, the horse raisers of the steppes combined it with a pugnacious ethos and a strong tribal command structure to become unbeatable. Their expansion transformed the social landscape of Eurasia. The valley civilisations could not prevail against chariotry. Sweeping conquests drastically changed the disposition of power and caused a moving and mixing of people. The realm of dominance of chariot tactics reached a limit in the heavy forests of western and northern Europe. Archery and chariots operate best on open ground. This lay to the east and rapid though indirect transmission spread the methods to the Yellow River by 1300 BC and into northern India with the Aryans between 1500 and 1200 BC. Conquest and a capitalistic form of warfare led to a concentration of surplus production, which not only sustained endemic violence, exploitation and brutality but also the arts of civilisation and a leisured class. This land based militarism had the effect of drying up the older sea links along the southern shores of Europe and Asia and replacing them with the transmission of political and social impulses through the constant, aggressive competition between culturally diverse peoples. This competition provided the driving force for an intermittant series of cultural developments geared to enhance the creation and deployment of power.

The balance between the steppes and civilisation was reasserted with the invention of iron smelting around 1200 BC. This provided infantry armour capable of turning the firepower of aristocratic charioteers. Barbaric, iron-using tribesmen overturned the chariot elites of Hittite Anatolia and Mycenean Greece. Egypt, devoid of iron ore, was deprived of its Palestinian and Syrian empire by the iron-using Sea Peoples by 1165 BC. The Assyrians had iron at hand and were able to arm the greater numbers of their sedentary population to turn away the attacks of iron armed tribesmen. The highly organised, massed infantry of Assyria provided the model copied by Persia, Macedonia and Rome. Beyond Assyria, iron had a levelling effect on war and power, disbanding the empires of mounted aristocracy in favour of tribal states comprising free peasants and small landlords, such as Israel. The aggressive spirit and dispersal of power in this arrangement was inherently unstable and the centralisation of Assyria prevailed over it and provided a heritage for Persia.

The Greek hoplite moving in unison in the close order of the phalanx raised the effectiveness of infantry to a new height. The social consequence of this development was quite profound, injecting an egalitarian, collective spirit into politics. This tendency was reinforced and democratised by the introduction of the trireme. Manning the rowing benches of the fleet, which made Athens mistress of the eastern Mediterranean, called for no more than muscle. Involvement in wielding power extended below the ranks of the citizen farmers to the urban demos. Changes in the equipment of war fostered a change in the social structure from domination by a mounted aristocracy to a more democratic distribution of influence.

Around 900 BC, a thousand years after the first appearance of the chariot, the military balance between the tribes of the steppe and civilised peoples was once more upset in favour of the barbarians. The cause was the novelty of fighting from horseback. The ability to cover ground at the gallop, travel faster and farther than any other kind of troops, climb mountain passes, cross rivers and penetrate forest with ease, gave a distinct edge to the grassland nomads. By 600 BC the Persians had conquered the Assyrian empire. The Scythians pushed south, the Celts west and the marcher lords of Ch'in adopted the barbaric methods to emerge as the rulers of China by 221 BC. A second wave of Mongols, Germans, Huns, Avars and Magyars swept out in the era 200-600 AD. The buffeting and jostling among these horsemen sent shock waves reverberating from one end of Eurasia to the other, creating an ecumene driven by conflict.

What re-established the balance against this light cavalry revolution was the development in the civilised world of heavy armoured cavalry.

This style of fighting appeared first in Parthia, south of the Caspian, around 100 BC. A race of big, strong horses was bred, capable of carrying enough protection to make horse and rider impervious to arrows. The heavy horseman could stand still and trade shot for shot with impunity while the steppesman dashed around, wore his animal out and emptied his quiver. At this juncture a charge by the heavies readily dispersed the light cavalry of the grasslands. The nomads could not adopt this mode of fighting because their pastures were not rich enough to sustain big horses. The support of these beasts was a considerable burden and, although they knew of the technique, the Chinese seem to have eschewed it in favour of carrying war to the tribesmen in their own fashion with light cavalry as a cheaper proposition. To the west, Sassanian Persia adopted this defence before 200 AD and began to evolve the social structure that went along with it. The beginnings of feudalism can be found in this means of military locomotion. To supply a formidable force of armoured cavalry distributed widely in the countryside, ever ready to mount a local defence against nomad attack, the horsemen were provided with an agricultural village apiece to administer to their needs. This mode of society and polity was adopted by Byzantium, where the name cataphracts was applied to the horsemen. From thence it spread slowly westwards, providing the institutional organisation which came to characterise Europe's Middle Age. By the 9th century Latin Christendom had adopted heavy cavalry and by the 10th century they had halted and turned the barbarian tide. In the wooded west the bow of the cataphract was replaced by the lance and emphasis was placed on the crushing shock of the charge. The concept of knighthood took shape among the Franks and the institution of the manor led to an effective decentralisation of power to a landed aristocracy.

At the same time clinker built boats provided Viking marauders with a vehicle for their piracy. As their raiding turned to trading the extent and depth of involvement in commerce and the urban foci of its trade routes in western Europe increased. Mastery of the high seas, by the 16th century, shifted the axis of European power to the Atlantic shore and planted the seed of European global domination.

The degeneration of Byzantium's cataphracts' fighting zeal made them easy prey for the disciplined, faithful certainty of the light cavalry that poured out of Arabia after 632 carrying Islam along the shores of the Mediterranean and Red Sea, through the fertile crescent and into the plateau of southwest Asia. After a brief retention of Arab tribal structure in garrisons in conquered lands, there was a reversion to the Iranian landed, military gentry model.

The steppes continued to generate waves of light cavalry with the Turks dominating Islam, Asia Minor and northern India by 1200. Genghis Khan directed the Mongol conquest of the steppe from the Volga to the Amur by 1227, and his successors extended their realm to eastern Europe, Persia, Mesopotamia, Anatolia and all of China by 1279, driving into the Punjab, Java and Burma in the 1290s, by which time the original driving force began to dissipate. The Mongols employed light mounted archers with a supplementary shock force of heavy cavalry. The methods of long distance communication and control employed in hunting were turned to military use in superior coordination of highly mobile columns, concentrating force at the right time and place. Point and flank scouting and fleet couriers provided the Mongol general with unparalleled control. After 1300 the Turkish drive picked up momentum and continued to expand into India and Byzantium reaching the Sava-Danube line by 1500.

What turned the tables against these horsemen was guns. The Chinese used gunpowder for military purposes around 1000 AD. It may have reached Europe with the Mongol invasion. It played a part in the wars of the Teutonic knights against Lithuanians and Russians, and Jan Žižka, the Bohemian leader, first employed artillery systematically in the Hussite wars of 1419-1434. In terms of the internal arrangements of Christendom, ordnance tolled the knell of knighthood and the degree of autonomy enjoyed by territorial magnates behind the stone walls of their castles. The wider disposition of power gave way to the centralised control of monarchies. Only the royal treasury could afford these complicated and costly weapons which could knock down fortresses as the *ultima ratio regnum*. In China, guns and Confucianism reinforced the central tendencies of the past, maintaining political unity from 1270 to 1912. The samurai of Japan contrived the survival of feudalism by excluding guns. When hand guns were developed to a reasonable level of effectiveness they reinforced the destruction of cavalry superiority which the longbow had begun. Although it was a long time before their rate of fire or accuracy was as good as the potential of the longbow, they required neither the strength nor skill necessary for the longbow's success. If guns destroyed Europe's own heavy cavalry's dominance they, by the same token, gave the infantry of the civilised world the means of breaking the charge of nomadic horsemen from the east. The manufacture of guns and ammunition required the organisation of mining, smelting, casting and chemistry which was neither within the capacity nor in tune with the inclinations of the nomads. After 1480 the Slavs of Muscovy turned the tide and drove through the horsemen across Asia with firearms and men produced by a densely occupied home base.

The cutting edge which made the globe a European sphere between 1500 and 1850 was driven by the bellicosity bred in the creative destruction of centuries of nearly constant feudal war. It was armed with vastly superior naval firepower. The Venetians first used mounted guns on ships in the 14th century. At the end of the 15th century ports were cut in ship's gunwales and broadside fire became the chief mode of operation, with the Portuguese leading the way in their battles in the Indian ocean. The steppe ecumene of Eurasia dwindled in significance after the surge of the Turks in 1507-1515. It was surpassed by the network of seaway links that drew the transactions of the world to a focus on the Atlantic face of Europe. In Europe itself, the nation state with centralised sovereignty was manifest in large standing armies, drilled to manoeuvre for the best deployment of firepower. The state's achievement of the monopoly of this elaborate and expensive form of organised violence, drew it into continuous contention with its foreign rivals. This national rivalry generated an upward spiral of technological advance seeking advantage in an intense and continuous competition. The organisation of the military provided the vehicle for the creation of the Prussian state between 1640 and 1740 and the military service nobility of Russia was created as an instrument of modernisation by the first Tsar in the 1560s and brought to a pitch of employment by Peter the Great after 1698.

With the dramatic advances in transport and communication of the 19th century, bringing the potential for speedy and extensive troop deployment and a broadened scope of control, there was call for more elaborate management and command organisations. The railway system of India was conceived of as a whole in 1853 with a view to administrative and military control. After the Prussians demonstrated their strategic efficacy in 1886, Germany's railways were laid down and controlled with military needs as the basic design criterion.

The integration of the social order and the industrial base into a national war machine was begun with the creation of the Prussian general staff engaged in the explicit activity of planning. The workings of the education system, manufacturing and the infrastructure of the economy were geared and articulated to produce material, train manpower and mobilise them for war.

The industrial base necessary to sustain a nation in warfare where the theatre is potentially global, limited the competition for hegemony to a few. In 1850 Britannia ruled the waves in keeping with her industrial dominance. From 1870, the industrial leadership began to shift to Germany and the USA and technological advance became applied science. In 1917 the USA joined the European war and the Bolsheviks overthrew the Tsarist regime and withdrew from the war. This marked the passing of global dominance from solely European hands. The Prussian general staff model was applied to the total operation of the state in Russia, Nazi Germany and Japan. German efforts to reassert this dominance fell to an alliance of the USA, UK, and Russia; the first two adopting the machinery of total mobilisation and the combined, planned operation of the economy and military sectors in the exigency of a fight for survival. The legacy of this exercise was a widespread view of government's competence and responsibility to control all the workings of society and eradicate the rubs of adjustment and external shocks to the fabric.

With Atlantic Europe quiescent and exhausted the USA and USSR turned on each other to contest direction of the world's fate. Their confrontation reached a stand-off created by the unspeakable capacity of both sides to transport enormous and lasting destructive power wherever on the globe it pleases in minutes. With the uneasy balance of mutual assured destruction holding the big battalions in check, since mid-century war has been local, sporadic and irregular. Although rocket borne nuclear weapons have turned the globe into an isotropic sphere where geodesics are great circles and surface conditions irrelevant, it remains the case that the only real victory lies in the control of people and territory, for which the surface must be invested and subjected. Thus, the texture of the land and the geography of war remains a vital political concern.

Readings

The epigraphs are taken from:
Sun Tzu *The Art of War*, translated and with an introduction by S.B. Griffith (Oxford University Press, Oxford, 1963)

There is a thorough coverage of the applications of geography to military affairs in:
L.C. Peltier and G.E. Pearcy *Military Geography* (Van Nostrand, Princeton, 1966)

The distribution of military power is treated in:
J. Soppelsa *Géographie des armaments* (Masson, Paris, 1980)

For the historical geography of military might we have drawn on:
W.H. McNeill *The Rise of the West* (University of Chicago Press, Chicago, 1963)

The most recent review of the study of violent conflict is:
F.A. Beer *Peace Against War: The Ecology of International Violence* (W.H. Freeman and Company, San Francisco, 1981)

War is placed in the general setting of international affairs in:
B. Russett and H. Starr *World Politics: The Menu for Choice* (W.H. Freeman and Company, San Francisco, 1981)

The nature of war in the second half of the 20th century is examined in:
Sir Robert Thompson (ed.) *War in Peace: An Analysis of Warfare since 1945* (Orbis, London, 1981)

2 INTELLIGENCE

> Now the elements of the art of war are first measurement of
> space; second estimation of quantities; third calculations;
> fourth comparisons and fifth chances of victory
>
> Sun Tzu, (chapter 4, verse 16).

Maps

The most potent instrument and symbol of intelligence, decision and
command in war is the map. From war rooms to cockpits, charthouses
and company headquarters, this is the means of recording and convey-
ing information, calculating moves and directing action. Emil Ludwig
wrote of Napoleon:

> Whether his halt be short or long, in wartime the map is always
> ready to his hand, in carriage or in tent, in camp or by the watch
> fire . . . Through all countries, for the whole duration of his life,
> the map follows him, pierced with coloured pins, illuminated at
> night by twenty or thirty candles, and a pair of compasses lying on
> it. This is the altar before which he offers up his prayers. It is the
> real home of the man who has no home.

If not universal, the skill of abstract representation of the lie of the
land and its occupants is widespread. The ability to build a generalised
picture of the surface of the earth in the mind and then convey this two
dimensionally to others seems to come quite naturally. Children take to
drawing maps quite readily. To coordinate fighting into concerted
warfare some knowledge of what lies beyond the immediately visible
landscape is necessary. The trace of a stick on the sand or more permanent marks have been used to bring things beyond eyesight to mind.
Scouts and spies have used them to describe unfamiliar terrain or the
enemy's whereabouts and leaders have used them to ponder and direct
action.

In an era when piracy, trade and politics were less discrete occupations, sailing charts had mixed military and commercial motives. If
the navigational needs of the Admiralty charted the seas so that
Britannia could rule the waves, the greater topographic information
required for moving and firing big guns inspired the precise surveying

of the land. It was Mercator's teacher, Frisius, who in 1533 provided gunners and geographers with their shared basic technique – triangulation. Sightings from two known points provide for fixing the position of a third distant point. Digges' invention of the theodolite in 1551 simplified the measurements and computations involved. Calculating ranges on the battlefield became a matter of science. To move men and trains of guns between battlefields still remained inexact, based on the vague representation of topography in 16th century maps. In England the active land market stirred up by Henry VIII's dissolution of the monastries reinforced the military demand for more accurate maps and the theodolite provided the means of making them. The cockpit of Europe, Flanders, was the setting for the earliest intensive mapping of a landscape for military purposes under the direction of the duc de Sully in the 1600s. The sorry state of geographical knowledge in Britain was dramatised by the panic of the Jacobite risings of 1715 and 1745. A survey of the Highlands was begun by the Royal Engineers in the aftermath but halted by war with France in 1755. In 1783 the engineers were given the task of triangulating southeast England to cooperate with the French in measuring the exact distance between Greenwich and Paris observatories to fix the position of their meridians. The French Revolution interrupted this work but by the same token provided the motive for extending triangulation and mapping to the entire country. Fear of invasion from France inspired the establishment of the Ordnance Survey in 1791 as a branch of the Royal Engineers. The Survey remained in the hands of the Royal Engineers till the 1960s. The title 'Ordnance' Survey indicates the initial objective of the mapping. To direct the fire of big guns with ranges beyond the horizon, it is necessary to have precise data on the location and altitude of guns and targets. Much of the development of survey technology has been a response to this particular need, increasing the scope and accuracy of the survey as the missiles reached further with more devastating results. The evolution from triangulation, through balloon observation, airphotography and radar to satellite sensors has paralleled the succession from black powder and smoothe bores, through cordite and rifling via bombers to guided rockets.

Besides the artilleryman, all branches need detailed map coverage of the ground and the ability to interpret topographic maps to guide their tactics. As if the significance of landscape detail in the fortunes of war needed emphasis, we can credit Napoleon's turning blunder at Waterloo to an error of topographic perception. His maps and local guides were not good enough to discern a sunken road cutting what appeared to be an open plain. The French cavalry were sent across the plain in pursuit of the retreating English only to come to grief as they piled on top of each other into this artificial declivity. The Ardennes have provided a recurrent occasion for geographic misperception and demonstration of the shortness of military memories. What is in fact readily traversible was viewed as a massive block and left as a weak joint in defenses, providing the opposition with a line of least resistance twice in the same war. It seems to be a rule that a physical barrier offers less friction to such an attack than does human opposition. Von Manstein and Guderian advised von Rundstedt to attack through the Ardennes in 1940, rather than repeat the attack through Belgium of 1914. In the closing months of World War I, the permeability of this dissected plateau had been revealed quite vividly. The Allies conceived of this as a back-stop against which a large part of the German armies could be trapped by advancing wings to abut against the eastern and western flanks of the upland. The major part of the German forces withdrew easily through the Ardennes at the cost of sacrificing a small rear guard. The Allies were to repeat this geographic mistake again and leave a thin screen to cover this part of their front in 1944 which von Manteuffel attacked in Germany's final counter-offensive. Deficient geographical foresight showed itself also in the Normandy invasion. Allied intelligence had failed to assess properly the penetrability of Norman bocage, and its small fenced fields reduced the progress of their armour to a crawl until a means was devised to push the hedgerows aside.

The foundation of topographic mapping services throughout Europe was military. In the USA, although the Army Corps of Engineers' surveying activities were undertaken to aid civilian navigation in the east, in the western territories surveys were taken as an aid to conquest and control. The economic motives embodied in the US Geological Survey only took over the mapping duty at the end of the 19th century.

At a broader scope, the planning and direction of wars rather than battles required reliable and informative maps covering a wider field at a smaller scale. The prominence and excellence of German cartography in this realm is not unrelated to the organisation of the Prussian staff. The superiority of this general staff in planning and executing war, demonstrated vividly in 1866 and 1870, depended on the accumulation and employment of a vast amount of geographic information on both large scale maps for tactics and small scale maps for strategy. It is noteworthy that the victor of 1866, von Moltke, studied under Ritter, the geographer.

The variety of maps required for doing battle has changed through time with the technology of war. To the end of the 19th century generals could usually see the entire battlefield that they fought over from a vantage point. Detailed topographic maps were not required for tactics at this level. Commanders could tell their subordinates where to manoeuvre by pointing to terrain features that they could both see. Stupidity and arrogance could distort the transmission of information, as when Captain Nolan's sarcastic indication of the valley of death was taken literally by Cardigan, who then led the Light Brigade to disaster. In principle, however, visual inspection and words and gesticulations were quite adequate. On open terrain even the likes of Wellington and Napoleon at Waterloo could physically see most of the battlefield. The principle need felt then was not for large scale topographic maps, but for small scale maps of the road network for commanding the movements of divisions of armies to a battlefield. Maps were accorded little tactical value. Even artillery was aimed at its targets by sight.

The increased lethality and range of guns in the 20th century led to the demand for accurate topographic sheets. The disposal of units to avoid presenting concentrated targets meant that after World War I commanders above company level could no longer physically see their battlefield and had to fight from maps. Battalion level manoeuvre is now directed by map coordinates. The use of long range, indirect artillery fire from World War I on also demanded accurate maps so that forward observers could call fire down on targets unseen by distant gun crews.

The increased reach and destructiveness of weapons through this century has induced not only an increase in troop dispersion but also a decrease in the appropriate map scale. In World War I company and battalion commanders used 1:25,000 scale maps. In World War II and Korea the standard small unit scale was 1:50,000. In Vietnam, airmobile company commanders were forced to use 1:250,000 scale maps because their helicopters flew across the expanse covered by a 1:50,000 sheet before they could figure out where they were. The dispersion forced by the use or threat of tactical nuclear weapons and the velocity of modern, mechanised warfare also encourage the use of small scale maps by company sized units.

The kind of terrain intelligence called for is a function of the scope of a military decision in space and time. Commanders of echelons above the divisional level, like the generals of the 19th century, are concerned mostly with lines of communication, with roads. They do not fight in a tactical sense but allocate and move units into or out of a battle. Corps and army commanders' decisions have a lead time of hours or days before they influence the course of a battle, because of the long times required to move major units even if they are motorised.

Tactical decisions at the divisional or lower levels, require more detailed and timely information. It is necessary to determine fields of fire for line-of-sight weapons; cover and concealment from enemy fire and observation; the ease of movement off the road for tracked vehicles; landmarks for helicopter routes and landing zones for helicopters. At this tactical level, decisions may have an impact within minutes. At the company level this may come down to a matter of seconds. In this setting the military aspects of the terrain may change as quickly as the disposition of enemy and friendly forces. An artillery strike may convert a village crossroads into a tank trap of rubble or a readily penetrable forest into an instant abatis. The passage of a tank company through a field may so churn the ground that it is impossible for other tanks to follow.

The increasing importance of terrain features and, thus, maps to mechanised military operations, has made the impact of erroneous or out-of-date maps potentially disastrous. Cultural features such as roads and the extent of the built-up area, may change dramatically between map survey and map printing. On a battlefield, bridges disappear, dams are breached and buildings reduced to rubble at a speed faster than the changes can be posted to the best of maps. In Vietnam, the US forces tended to use recent air-photos rather than maps for terrain analysis because of the rapidly changing face of the landscape in battle zones.

The results of following maps which are in error can have far-reaching and damaging effects. European Russia is about 40 per cent forested. The unreliable maps available to the Wehrmacht in World War II showed almost continuous forest. The pronouncements of the general staff on the eastern front were couched in terms of continuous forest, an impression gained from faulty maps, not from observation of the battleground itself. In the early days of the Vietnam conflict the South Vietnamese army used French maps while the US Air Force used US maps. The result was air strikes being called in on the wrong village or on friendly troops, until US maps were distributed to the Vietnamese so that everyone was working on the same basis.

Terrain Analysis

What you need to know about a landscape for fighting purposes includes everything on the surface of a potential or actual battlefield that may affect operations. You need to know how easy a type of terrain is to travel across in general and where there is easy or difficult passage in particular. Places where equipment and men may be landed by sea or air must be discovered. Sites for encampments, defences and route-ways must be established. Having located base points and lines of movements, actual engagement with the foe is concerned with sight lines for shooting, the protection and screening afforded and the radius of contact associated with a general type of terrain. In particular, specific places to defend or strike and to locate men and guns must be ascertained. These issues concern both the man-made and natural components of the landscape. A tank commander is indifferent as to whether a steep sided water course is natural or a canal dug by man. It is an obstacle to cross-country movement regardless of how it got there. The back slope of a hill, a pile of mine tailings or a stone farm-house all provide cover from direct fire for a tank, and all three are part of the same military terrain.

Given that the human contribution to the landscape has been quite superficial until recently, it is not surprising that much research on the terrain for military purposes has been related to physical geography. The scale, attention to detail and measurements involved in answering military questions are of a similar level of resolution to those of geomorphologists concerned with the processes and structures which generate landforms. The emphasis on certain types of features, terrain and environmental setting in the literature of physical geography bears a striking resemblance to scenes of battle past and potential. In some cases it seems that existing lines of inquiry were fostered because of their coincidental relevance for war. The interest in desert landforms exemplifies this sequence. In other cases research which was encouraged because of its possible military significance has proven academically fruitful but insignificant militarily. The intensive research on slope development in the 1950s was funded by the US Army because of its possible relevance for tank operations, but it produced no startling results from this viewpoint. It is possible that military attention to certain aspects of the landscape provided the inspiration for academic interest. There certainly has been a good deal of research in physical geography commissioned directly to serve the more efficient prosecution of war.

One fascinating case of this interaction between physical geography and fighting provided the basis for the Long Range Desert Group in the Western Desert. During the 1930s R.A. Bagnold, a Cambridge geographer, did extensive field work in the Libyan Desert, accompanied by a number of others. They explored the varieties of terrain formed by windblown sand dunes, barkhans and sandsheets. To navigate the desert they devised the sun-compass. With the outbreak of war in North Africa, Bagnold organised the Long Range Desert Group to exploit this experience in a most effective means of disrupting Rommel's interior lines and destroying aircraft on the ground.

In 1918 the French army had produced maps showing where tanks could pass. Following this precedent in North Africa the Royal Engineers mapped the friction that varieties of desert terrain offered to wheeled or tracked vehicles over the fighting ground. These 'goings' maps present the surface differentiated according to the ease of movement it afforded. Maps of this kind were also produced for the Anzio beachhead. Since 1945 the British and American army engineers have pursued the analysis of desert landscape for war using the Trucial States and the vicinity of Yuma as their respective field study areas. Academic geomorphologists have been actively engaged in these endeavours. Interest in war in a desert setting was sustained by the Arab-Israeli conflicts and the location of oil resources and is hardly likely to wane in the years of the 'arc of crisis'.

A major concern during World War II was the amenability of sections of coast to landing in the face of opposition from the shore. This was a crucial issue in planning the Normandy invasion and the stepping-stone campaign from island to island of the Pacific. To judge how readily and speedily men and vehicles could be landed from the sea, the slope of the beach above and below the water and the load bearing properties of its composition were critical. The Baker Street Irregulars of the SOE (Special Operations Executive) collected picture postcards and holiday snapshots of France's northwest coast and even landed men from submarines to take samples of beach sand and pebbles to acquire the necessary information. A group of geographers and surveyors under the leadership of W.W. Williams of Cambridge found means of determining the way a beach fell off below the water line from measurements of the spacing of wave crests on air photos. The American Corps of Engineers and the OSS (Office of Strategic Services) had topographic intelligence sections which made scale models of landing sites and other targets for attack using air photo data.

Since World War II efforts have been made not merely to describe conditions in surveyed localities, but to generalise surface conditions so that characteristics of the terrain can be predicted from remotely sensed images — air photographs or satellite signals. The object is to be able to draw a description based on known ground from a file organised in categories of terrain which can be recognised on air photographs. Separate sets of categories can be established for different climatic conditions. Given an air photo of a particular part of the world to which troops are to be dispatched, it would be desirable to be able to predict relevant surface conditions by matching the characteristics of parts of the photo to the file categories. The product would be a map of landscape types. British Army engineers, soil scientists and geographers tested the idea in temperate, arid, savanna and tropical settings with survey data gathered in Britain, North Africa and the Middle East, East Africa and Malaya, concentrating on the home ground around Oxford for logistic ease, and Libya and the Trucial States as examples of the strategically most significant desert theatres. The conclusion of these experiments was that a militarily useful system of predicting ground conditions at a very fine resolution is feasible. Although it has not been pursued, such a method of providing intelligence could be put into operation very easily and rapidly.

Rather than producing a fine-grained regionalisation of any piece of territory in terms of descriptions of physiology and geology, the American and Canadian armies have gone for quantitative measures of the lie of the land as they bear on sight lines and how good the going is. The US Army Quartermaster Corps' researchers took measurements of the spacing between hills and valleys, the difference between highest and lowest points, the average slope and the frequency of changes from up to down slopes over a unit area. It was found that a wide array of samples of the terrain of the USA fell into 25 basic categories. These 25 classes captured the major varieties of landscape in a parsimonious fashion.

The Corps of Engineers concentrated their attention on desert terrain using test sites in Arizona. They sought the best set of characteristics of the terrain to measure and map for military purposes, principally for determining 'trafficability' — the ease of passage for vehicles off roads. The landscape was analysed in terms of a number of measures of surface geometry, geology, soil and vegetation. Generalised landscape types were composed from measurements of slope, relief, layout and cross section of the terrain. To assess qualities for military use, pertinent characteristics can be put together into a composite map. Trafficability, for example, would be portrayed by the overlaying of maps of local bearing strength of the soil and the average inclination of slopes. The Canadian Army engaged in similar efforts in mid-latitude conditions, mapping surface characteristics from air photo data to predict the speed vehicles can achieve cross country.

On the face of it the advent of guided missiles released the exchange of firepower from its earthbound subjection to the friction presented by the face of the land and sea. With rockets that can traverse oceans and continents, the best routes for the delivery of destruction become great circles and the earth can be regarded as an isotropic spheroid. Even some missile systems, however, require an appreciation of terrain. One of the chief recruiters of cartographic talent over the last few years has been the US Department of Defense, intent on devising the guidance system for the ground hugging cruise missile, which involves negotiating the ups and downs of the surface.

Beyond terrain appreciation, the effect of the variety of geographical settings on men, weapons, vehicles and tactics has inspired close scrutiny. When engineers cost road construction work in the tropics they still refer to data on work rates collected by the British Army in India. What are unusual conditions to temperate lowlanders have been special targets. Mountains, forests, deserts, snowfields and jungle have been examined in both historical works and analytic works making preparation for future action. The design of vehicles, equipment, weapons and operational procedures should obviously pay careful attention to the stresses, limitations and opportunities offered by the environment. Geographers have been closely involved in the Quartermaster Corps' research on means of combating cold and heat, damp and dryness, dust and mud, wind, and dense vegetation. Amongst other things the colour of equipment must be matched with the setting to provide the best camouflage. Olive drab may not merge too well with the sandy desert in which several possible theatres of action lie. The details of some of these matters of adaptation to the physical environment will be discussed when we turn to tactics.

If a knowledge of climate and its effects plays a part in the preparation for action, the more local issue of weather has a crucial bearing on the timing of action. It would be improper to tag meteorology as a mere branch of geography, yet the predictions of weather necessary for fighting purposes are geographical. They involve particular places as well as the expected time at which certain conditions will prevail. What finally tipped the balance of the decision to go with 21st June for D-Day was a meteorological judgement that the weather would remain calm in the Channel for long enough to establish a beachhead. The timing of the Rhine crossing hung on a prediction of foggy conditions. In cunning perversity Slim launched his counter attack in Burma in the monsoon's highly predictable deluge, when action was least expected and Japanese defences were lulled into somnolence.

Those who fight at sea or in the air have continuously tried to reduce the chance involved by improving their ability to foresee weather or obviate its significance by improving the performance of ships and planes against the buffeting of wind and water and developing the means of sensing the enemy's presence through fog and cloud. Remote sensing serves the broader purpose, of course, of seeing beyond the horizon.

Much of the experience of environmental extremes and the observations and measurements which have excited scientific inquisitiveness was undertaken for a military purpose in the first place. In some notable instances wartime operations brought to light otherwise unobserved phenomena which have enlightened our understanding of the world in which we live. Fliers on bombing raids to Japan in 1944 discovered a narrow stream of air around 30,000 feet above sea level, travelling at over 200 mph in the forties of latitude. They used this to assist their return journeys. The frantic efforts to capture Iwo Jima were occasioned by the need for an advanced base to provide succour for bombers which had used fuel heavily in doing battle against this headwind in their westward flight. This experience of what came to be called the jet stream, provided a link in the growing understanding of atmospheric circulation. This is not to say that war is the best or only forcing ground for knowledge, but in forcing people to extreme measures it has sometimes broadened their experience and vision.

Military Education

The most direct contribution of geographers to the preparation for war is the education of soldiers. They have had a traditional role in teaching an appreciation of landscape and map reading. Positional sense, for the individual soldier as for the ball player, is a basic skill which comes naturally and is enhanced by experience. Field training serves as the military equivalent of match play in developing territorial feel and foresight. Above and beyond the individual scope, the task of command from the squad leader up to the supremo, calls for an ability to sense space in the abstract via a map and analyse positions and moves on this simplified representation of the land. This requires what von Clausewitz called 'Ortsinn' — a sense of place:

the power of quickly forming a correct geometrical idea of any portion of the country, and consequently of being able to find one's place in it exactly at any time. This is plainly an act of imagination. The perception no doubt is formed partly by means of the physical eye, partly by the mind which fills up what is wanting with ideas derived from knowledge and experience, and out of the fragments visible to the physical eye forms a whole; but that this whole should present itself vividly to the reason, should become a picture, a mentally drawn map . . . all that can only be effected by the mental faculty which we call imagination.

It is natural that the scope for the exercise of this talent should increase along with rank. If the hussar and rifleman in charge of a patrol must know well all the highways and byways . . . the chief of an army must make himself familiar with the general geographical features of a province and of a country; must always have vividly before his eyes the direction of the roads, rivers and hills, without at the same time being able to dispense with the narrower sense of locality.

The geographer's part in cultivating a superior capacity to exploit ground among soldiers and the evolution of this function can be traced through the fortunes of geography at West Point. One of the first two instructors at the Military Academy was appointed in 1803 to teach field sketching. A geography department grew up as drawing was enhanced with cartography, surveying, geology and engineering fundamentals. The strong emphasis on maps and engineering led to the department being renamed Military Topography and Graphics in 1942. The need for a broader perception of the world was met by introducing courses on cultural and political geography. The scramble to catch up with the Soviets after their Sputnik success made aerospace the new touchstone and in 1962 the title was changed to Earth, Space and Graphic Sciences. The responsibility of the department for teaching cadets how to use slide rules, how to set about solving engineering problems and how to do engineering drawings, led quite naturally to it taking on the task of teaching the use of computers. The major growth in curricular offerings was in the fields of geography and computer science and so in 1981 the instructional unit was entitled the Department of Geography and Computer Science. The department teaches terrain analysis to all second year cadets at the Academy and geography is available as a field of concentration. The geographical and computational components of the department blend in its computer graphics laboratory over the problems of information systems management and automated cartography. This educational structure is a clear recognition that geographical intelligence will in the future be fed to all levels of command by computing machines.

Accompanying the increasing emphasis on automated intelligence there has been a greater awareness of the value of using the ground to advantage. The dramatic improvement of weapon accuracy has raised the probability of a first round hit and placed a premium on cover. In 1945 it took thirteen rounds for a tank to have an even chance of hitting a stationary target at a range of a mile. By 1975 the first shot had a fifty-fifty chance of scoring a hit. The Yom Kippur War provided ample evidence of this lethal improvement. On the strategic scene NATO saw itself faced with the prospect of doing conventional battle against the superior numbers and improving equipment of the Warsaw Pact. One means of offsetting these disadvantages was seen to be superior exploitation of the terrain.

The current doctrine of the US Army expressed in its operations field manual (FM 100-5) makes a good deal of the use of the advantages of the lie of the land. Instructions as to the objectives of terrain evaluation are contained in a hierarchy of manuals: for intelligence officers who provide geographic information (FM 30-5, FM 30-10); the battalion commander (FM 71-2); the company commander (FM 71-1); and platoon members (FM 7-7) in tank and mechanised infantry outfits. The principal components of the landscape which soldiers are admonished to seek are:

> points of observation
> fields of fire
> cover and concealment
> obstacles
> key terrain
> avenues of approach

The objectives of analysis are, thus, fairly clearly defined. Experts have enunciated systematic procedures for working out the best deployment to take tactical advantage of the terrain for attack or defence. What has not been contrived is a way to ensure that different soldiers produce the same best deployment faced with the same maps and enemy. The stumbling block lies in the landscape and map reading. Beyond the elementary skills of matching places on the ground with places on the map there is a need to train soldiers so that they are capable of forming a consistent mental image of the landscape from their visual impression of the map.

Misinformation and Propaganda

As well as aiding in the task of fighting, maps and geographical lies
have been used to mislead the enemy or gather popular support for a
dubious cause. The German school of geopolitics excelled in the cartographic portrayal of the menace to the Reich of those it wished to
conquer. Thrusting arrows expressed threats of encirclement. Boundary
lines were imbued with aggressive quality. Maps published in 1934
showed most of the Reich within range of bombers flying from a
continuum of points along the Czechoslovak border, and exaggerated
the threat of the Bohemian 'fortress' by misrepresenting the relief of
the region, showing Bohemia and Moravia as a block the same height as
the Alps. These maps published in *Zeitschrift zür Geopolitik* were
intended mostly for the consumption of journalists, teachers, and
foreign opinion formers. The journal was, however, put out as cheaply
as possible to achieve a wide circulation from the newsstands and
written for a lay readership. When the Nazis came to power they
appointed Kurt Vowinckel, the publisher of *Zeitschrift*, head of the
German publishers' association, commanding the output of print in
the Reich. As the geopolitical theorist Rupert von Schumacher put it
'Every political map is a weapon'. The influence of geopolitics had
permeated school geography long before 1933 and with Hitler's ascend-
ency 'patriotic geography' and 'defence geography' became the order of
the day. Atlases contained maps of Germanism spreading deep into
Russia and depicted political thrusts with spearheads. *Time* magazine
cartography is, of course, not above this kind of conscious bias to stress
a political point.

    One delightful example of cartographic subterfuge involves the
'goings' maps of potential mobility produced by the surveyors of the
Royal Engineers in Cairo during World War II. They produced a bogus
version of these maps and allowed them to be captured by the Afrika
Korps. This deception did succeed at least once in directing a large
formation of German tanks into impossible ground. In similar vein the
public announcements of the places where German rockets were
landing in 1944 were deliberately misplaced to the east, so that the
mean point of impact of the scatter which the Germans were targeting
on was shifted two miles east a week down river away from Central
London into rural Essex. Geographic misinformation can be a valuable

defensive tool.

Readings

The quotation on Napoleon and maps is from:
E. Ludwig *Napoleon* (Boni and Liveright, New York, 1926) p. 336
and that of von Clausewitz is taken from:
J.I. Greene *The Living Thoughts of Clausewitz* (Longman, Green and Co., New York, 1943), p. 25

Cartography and military technology is discussed in:
G.R. Crone *Maps and Their Makers* (Hutchinson's University Library, London, 1953)

Military terrain analysis is treated in Chapter 10 of:
C. Mitchell *Terrain Evaluation* (Longman, London, 1973)

The US Army Field Manuals (all published by US Government Printing Office, Washington, DC) referred to are:
FM 7-7, The Mechanized Infantry Platoon/Squad (1977)
FM 30-5, Combat Intelligence (1973)
FM 30-10 Military Geographic Intelligence (Terrain) (1972)
FM 71-1 The Tank and Mechanized Infantry Company Team (1977)
FM 71-2 The Tank and Mechanized Infantry Battalion Task Force (1977)
FM 100-5, Operations (1976)

The examples of geographic and cartographic deceit come from:
W. Stevenson *A Man Called Intrepid: The Secret War 1939-45* (Macmillan, London, 1976)
D. Whittlesey *German Strategy of World Conquest* (Farrar and Rinehart, New York, 1942)
and from personal correspondence with R.F. Peel, who also provided much of the material on geomorphologists and the military as well as being a source of inspiration.

The transformation from geographer to soldier is exemplified in:
R.A. Bagnold 'Early Days of the Long Range Desert Group' *Geographical Journal*, vol. 105, no. 1, (1945), pp. 30-46

Another stimulant for this chapter was a master's thesis in geography from Florida State University:
J.B. Green *Military Geography: Tactical Terrain Analysis* (1979)

3 LOGISTICS

> Pay heed to nourish the troops; do not unnecessarily fatigue them
>
> Sun Tzu, (chapter 11, verse 32).

Mathematics and Military Decisions

Once a commander knows where he has to go and where the opposition is, the next problem is whether he can feed, arm and replenish his forces sufficiently well to engage and overcome the enemy. An army marches on its stomach and gas tanks and empties its magazines at an alarming rate in combat. It was once possible for a commander to plan operations personally and direct a commissary and quartermaster to supply the necessary provisions. With the rise of industrial nation states and mass conscript armies with complex weapons, communications and transport, offering the prospect of continental conflict, the coordination of the economy's production and the needs of the military exploded the planning problem in space and complexity far beyond the individual's ken. In the 1860s the Prussian Army created a general staff of specialists supplementing the commander in making detailed planning and assignment decisions. This institution was adopted by the other great powers.

The grind of World War I and its enormous attrition in men and material occasioned the subdivision of planning among a number of staff agencies. This generated the needs for the facility to dovetail the vast array of plans and requirements and coordinate them with the civilian economy in an era of total mobilisation. The size and complexity of World War II lead to a gigantic increase in the planning function on both the military and civilian sides. The emergence of air forces to positions of supremacy with very elaborate weapons and supply needs and little in the way of institutional precedents, provided the setting for the application of mathematical programming to managing the war machine. In 1943 the US Air Staff established a programme monitor to compute the appropriate deployment of units to combat theatres, training requirements for flyers and technicians, supplies and maintenance needs to satisfy a plan devised to attain certain war objectives.

The work of economists representing the interactions of the economy as a system of simultaneous equations converged with mathematical developments in the solution of such systems and the construction of machines that could do the necessary calculations rapidly enough to be effective aids to choice and decision.

The Transportation Problem

In the vanguard of this enterprise one peculiarly geographic and fundamental logistic problem was solved before the general class of such programming problems. This was the transportation problem. This is the problem of a dispatcher dealing with a large fleet of vehicles shipping material from a number of supply dumps to a large number of points of need. The problem is the same at the global level, where the vehicles are tankers and the supply points are oil refineries and the demand points are theatres of war. This is the basic logistic problem, needing to be solved continuously at a variety of scales of operation. The objective is to consign vehicles to tasks in the most efficient manner, to carry out the transport at the least cost. The same task confronts civilian transport managers and schedulers daily and perennially, to be solved afresh as the geography of demand and supply changes in response to a multiplicity of unforeseeable circumstances.

To generalise, the problem involves meeting the given needs of a set of demand locations from the given quantities at supply points where the unit cost of shipping from each supply point to each demand point is known and does not change with the volume shipped. The solution will be the pattern of flows between points of supply and demand which keeps the cost of transport as low as possible. The transport cost is calculated as the quantity shipped times the unit cost summed up over all flows. This everyday, practical exercise can be cast as a mathematical problem. Having expressed the problem in general, formal terms it is possible to prove that a unique, best solution exists for any particular problem of this kind. In that knowledge a common method can be devised, guaranteed to find the best solution to any specific instance of the problem by a series of repeatable calculations. The object and the conditions that circumscribe the possible solutions to the problem can be represented by a set of linear equations and inequalities. The computational rubric, the algorithm, is then brought to bear to find the values of the unknown variables, the flows, which solves this set of expressions simultaneously. In geometric terms the system can be conceived of as a multi-faceted dome, a convex polyhedron, with the possible solutions lying at the junctions where the lines defining the facets cross. The premier algorithm used moves from junction to junction in search of the best solution. The method of solving transportation problems of large dimensions by such means with a computer was perfected by 1950. The next year the method was extended to a variety of non-spatial, programming problems and was rapidly applied to solve industrial engineering allocation and design problems in oil refining, chemical, iron and steel and power production, besides the obvious applications of the transportation algorithm in the civil transport and communications sectors.

Thus the convergence of insight gained from learning by doing in logistic scheduling with mathematicians' abstractions and the emergence of electronics led to a method for invariably finding the best answers to a wide variety of military and then civilian logistic planning problems. It is significant that slightly prior and independent developments in mathematics were promulgated in Russia in connection with the central planning of freight operations. This was not followed up and did not result in operational methods. It is evident, however, that this was a problem whose time was ripe for solution. The generalised form of the problem, besides its civilian employment, was applied not only to spatial, military problems such as deployment and air life routing, but also to a variety of non-spatial problems such as contract bidding and personnel assignment.

The practical transport problem is largely one of moving fuel and ammunition, which account for over 90 per cent of the tonnage of logistic support required by a modern mechanised army. Food, medical supplies, spare parts and the like are a small part of the total volume to be transported. The main logistic worry is how to get diesel fuel, gasoline and artillery rounds forward. The balance between fuel and ammunition varies. Static, defensive operations use far more ammunition than fuel, while an exploitive attack, such as that of Patton in northern France or Sharon on the east bank of the Suez canal, will burn up a greater tonnage of fuel than of ammunition.

Increasingly, the movement of fuel and ammunition is being accomplished with 'minimum break of bulk' techniques. In Vietnam, for example, a driver and truck would load up with a truck full of 105 howitzer rounds directly from a ship in the port of Saigon. First Logistical Command would then order the driver to take them directly to a firing battery somewhere in Vietnam. Depots and intermediate transfer points were cut out entirely. The advantage of this procedure is that it cuts down tremendously on the number of people required to move supplies forward, since unloading and reloading at intermediate supply dumps is eliminated. This has enabled armies to increase their 'teeth to tail ratio'. To do so it places a premium on computational capacity. Logistics becomes extremely dependent upon computer assisted allocation and dispatching systems and reliable communications. Flexibility is purchased by spending more on calculations and information. US Army doctrine now calls for redundant back-up computer capacity at corps and theatre logistic command levels.

Systems Analysis

Many of the day-to-day, operational problems of getting fuel, ammunition vehicles, food and men to the right places, then, were no longer matters of guesswork. Dispatchers' rules of thumb and experience are supplanted by precise solution. The same mode of thought and variety of methods applied to solving operational problems associated with war plans arising from general staff intuitions, were directed to the strategic questions themselves. With the age of nuclear bombs, long distance bombers and missiles, came the need to evaluate global strategies and design intercontinental weapon systems. Given the technology, the underlying questions are, again, geographical. They concern both where in general and where in particular the elements of a system should be located to achieve a geographically specific mission. As an example of the analytic approach to military decisions consider the selection and operation of strategic air bases in the 1950s by the US Air Force. The problem was where and how to base the long range bombing mission aimed at the Soviet Union and how to operate this force in conjunction with the bases chosen.

The technological circumstances which presented this design problem partly arose from scientific breakthroughs and were partly a response to evolving geopolitical perceptions. The genesis of the USAF's search for the means of delivering bombs across oceans lay in the fall of France in April 1941 and the consolidation of Germany's hold on Western Europe. The B-36 was conceived of as insurance against the fall of England, a bomber with a 10,000 mile range capable of delivering 10,000 lbs of high explosives. Questions of base locations, possible targets, cruise and maximum speed requirements and ceiling altitude in the face of enemy defences all came to bear on the design. All of this was done in innocence of the development of the A-bomb until it dropped in Hiroshima in April 1945. In the course of building prototypes the strategic picture changed radically. By mid 1943 it was evident that Britain would stand but a Pacific war was in progress. In the war with Japan, because it was not clear that the stepping-stone campaign of retrieving Pacific Islands would succeed, there might be a role there for an intercontinental bomber. As the Japanese were pushed back and the war in Europe drew to a close, the immediate need for a B-36 began to evaporate. The Air Force, however, was starting to think in terms of World War III with Russia as the enemy. With uncertainty about the prospects of obtaining overseas bases and the extended base to target radii this implied, the B-36 looked like a good bet. At that juncture, strategic bombing with high explosives was only a marginal component of warfare. With the revelation of the destructive power of the A-bomb, the development of a bomber and ancillary system capable of devastating Russia from across the ocean was raised from a marginal to a central issue of strategy.

In 1947 strategic bombing was still conceived of in World War II terms. The mission of a bomber force was firstly to penetrate enemy defences; secondly, reach and find the target and then destroy it and, finally, to return to base. The casualties that could be borne in getting to and from the targets depended upon the proficiency of destruction. The improvement of jet interceptors and anti-aircraft missiles made getting there and back more difficult. At the same time, nuclear bombs enormously increased destructive capacity. Building the potential for annihilating Russia's industrial and administrative centres was seen as a way of avoiding the costs of attrition of conventional war. The alternative to the slow but long legged B-36 was to combine the speedier B-29 with tankers for air refuelling to increase penetration of jet fighter defences.

By 1950 the Strategic Air Command had a force of B-29s and B-36s and B-50s and its role was seen as deterring a Soviet attack on Western Europe with the US monopoly on atomic weapons. If the US was called out, the SAC objective was to destroy the Soviet industrial base for waging war. At the time it was presumed that if the Russians came up with nuclear weapons, they could not produce many and would have difficulty delivering them since their long-range flying capacity was severely limited.

In 1951 the Air Force commissioned the RAND corporation to study the selection of overseas air bases to meet SAC's task. The analysts pointed out that the choice of bases was crucial in determining the make-up, destructive power and cost of the entire strategic force. The costs of acquiring, building, maintaining and operating bases alone was not sufficient criterion to employ in selecting bases. The geography of bases affected the costs of extending the range of planes which could not reach targets without refuelling. It affected the routes bombers had to fly through enemy territory and, thus, their potential losses *en route*. The vulnerability to attack and, thus, the resources necessary to keep bases operational and the costs of disrupted service, varied with location. The question was not merely where to locate bases but what system of bases and operations would best achieve the war mission of SAC.

Taking what was in the pipe-line in the way of aircraft and base operations for 1956 as a starting point, the analysts sought the arrangement of bases which would either enable the destruction of a given number of targets at least cost, or destroy the largest number of targets with a fixed budget. Which was the least cost or most effective system would clearly depend on a wide range of future variations in objectives and technical capabilities. Given the uncertainties involved, the analysts sought to specify the system which performed well under a range of possibilities concerning enemy offensive and defensive capacity and international political developments.

The existing system consisted of peacetime US bases for all aircraft, which would be moved up to overseas bases in wartime, except for some heavy bombers which would only use overseas bases for staging. The options considered were of four general kinds. Firstly, overseas wartime bases close to the enemy; secondly, wartime bases overseas but at some remove from Russia; thirdly, bases in the US with air refuelling; and fourthly, US bases with refuelling at overseas staging points, were compared.

The crucial variable in comparing these arrangements was distance. The systems differed in the distance from bases to four critical points. The location costs of the systems varied according to the distance from bases to targets, to the best point of entry to the enemy defences, to refuelling points and to points from which the enemy could attack bases. The configurations characterised by these distances determined bomber radius and logistics costs. They affected likely enemy defence deployment and, thus, the number of bombers lost in flight. They affected the vulnerability of bases and bomber losses on the ground.

As well as these 'locational' costs arising from the spatial disposition of the system, *in situ* 'locality' costs, which varied geographically with climate, terrain, construction costs, and in the extent of existing air defences were examined and incorporated in the analysis.

Very early in the process of examining the cost of increasing flight radius, the cost of penetrating the defence, the cost of bases outside the USA (including locality costs) and the cost of base vulnerability, the ground-refuelling option showed its superiority. It was superior in logistics cost terms and in the protection it afforded from attack. The comparison proceeded with a conscious bias against this sytem's performance, in effect trying to refute its superiority to be sure of it. Under a variety of possible future campaign conditions, including Soviet possession of atomic weapons, this was the most adaptable, robust solution.

What had started as a matter of logistics turned into an exercise in grand strategy. The report presented in 1953, indicating the weaknesses of SAC in the face of surprise attack, did influence Air Force policy to cut back the functions of overseas bases, harden its bases against attack, institute an airborne alert and develop long endurance aircraft. The general lesson which can be drawn from it is a reinforcement of the jingle about the loss of a war through want of a horseshoe nail. To have a chance of success, grand strategies must be built up on sound logistics and a thorough understanding of the spatial relationships involved.

More recently deliberations over whether and how to deploy the MX missile in the USA have been subject to a confused mixture of military, logistic and local political motives along with geopolitical perceptions. In June 1979 President Carter gave the go-ahead for the development of the biggest MIRV (multiple warheaded missile) allowed by the SALT II agreement. Its main features were to be the power and accuracy to destroy hardened Russian missile silos. It was a concession to the hawkish elements who had latched onto James Schlesinger's doctrine of 'counterforce capability'. This prescribed the acquisition of weapons to match or outweigh every component of the Soviet deterrent. Schlesinger had initiated the design studies for a mobile ICBM (intercontinental ballistic missile) in 1974. With this offering Carter hoped to reconcile the Senate to ratification of the SALT II treaty.

Viewed in purely military terms the ability to bust hardened silos makes this a weapon with which to hit first. It only makes sense if it is to be used to wipe out Soviet ICBMs on the ground. It was officially referred to as having a 'second strike ICBM countersilo capability,' because it is contrary to proclaimed US policy to launch a preemptive first strike. Yet in 1977, in the course of SALT II negotiations, when the MX was being treated as a bargaining chip, Carter's security advisor Brzezinski described it as a first threat against their rockets. In 1979, however, when the MX became a pawn in the internal politics of defence, it was called a second strike weapon and Brzezinski described it as providing 'the ability to respond in kind'. It is noteworthy that in April 1982 four American elder statesmen, McGeorge Bundy, George Kennan, Robert McNamara and Gerard Smith felt compelled to implore the US administration to declare that it would not use nuclear weapons first. The US continues to employ the threat of first use of 'battlefield' nuclear weapons as a response to Soviet conventional aggression.

Whatever the words used, the ability to smash through reinforced concrete shields and destroy missiles in their silos is effective only if it is used first. Used in retaliation against a Russian first blow, it will merely destroy empty silos. Once a first strike has been launched and some of the targeted opposition ICBMs are expected to survive and be used in retaliation, the mad logic of this game suggests unloading everything you have at every target in sight. There is indeed a fifteen minute warning of any retaliating wave of missiles within which to execute this. To have the ability to smash silos in retaliation is no deterrent to a potential aggressor. The USSR has no incentive to hit first unless they were confident of knocking out all US missiles. If they hit all US missiles it would not matter whether they were silo busters or not. If some US weapons survived, it would suit the Soviets to have them pointing at empty silos. The assured devastation of Soviet cities which the other two legs of the US triad, submarine and bomber based missiles, could wreak, would, however, make a Soviet first strike suicidal. If, in the light of this, the USSR sees the US deploying $100,000m worth of the silo busters even though its overall deterrent still enjoys superiority, what are they to think other than that they are being installed in preparation for a preemptive strike against Soviet land-based ICBMs? These account for three quarters of the Soviet force, as opposed to one quarter of the US nuclear armoury. In the face of this the Soviets might be tempted to hair trigger their missile force in preparation to launch a preemptive strike. They will certainly be inclined to seek a means of reducing their vulnerability by deploying more silo busters in a mobile, multiple-point basing scheme, leading to a further spiralling of the arms race and the dangers of perceived imbalance.

To fulfill Schlesinger's misperceived need for counterforce capability, the MX system has to be deployed in such a fashion as to have some striking power left after an attack by Soviet SS18s and SS19s with multiple warheads. The design criterion for the deployment in the first instance was determined by the killing capacity and maximum number of Soviet warheads which were being negotiated in SALT II. After an hypothetical Soviet first strike, it was calculated that 1,000

warheads would be needed to destroy any Soviet missiles left in launchers. This set the size of the system at 200 missiles with 2,000 warheads because only half the warheads would survive a Soviet attack
with the missiles it was entitled to under SALT II. (Coincidentally
2,000 warheads would make a mess of the Soviet's 1400 ICBMs in a
first strike.) To achieve the survival of half the force, this big, land-
based missile had to be provided with a means of protection. A variety
of geographical shell games with 200 missiles being shuttled between
4,600 launching shelters was investigated. The Air Force first determined that the best location was in the western plains. This met with
opposition from Kansas and Nebraska agricultural interests. By 1979
the Carter administration had settled on eastern Nevada and eastern
Utah. The logistic support for this weapon system in terms of its
requirements for 6,000 square miles of land, enormous amounts of
water, cement, earth shifting and construction workers, and the disruption of the region's social and environmental equilibrium generated
powerful opposition. Senator Laxalt from Nevada is a close associate
of Reagan. The alternative of placing some of the system in west Texas
and New Mexico met opposition from Vice President Bush and the
chairman of the Senate Armed Services Committee, Senator John
Towers of Texas. It came as no great surprise when, in October 1981,
Reagan announced that the MX would be housed in converted and
'super hardened' Minutemen silos in North and South Dakota, Montana,
Wyoming, Nebraska and Colorado. This was rejected by Congress and
at the time of writing the Senate Armed Services Committee is refusing
to appropriate funds for the MX until a permanent deployment decision is made. The options being considered are a denser version of
the shell game on military land (Dense Pack); emplacement 3000 feet
deep in mountainsides (Hard Tunnel); airborne launchers (Big Bird)
and Orbital Basing, involving putting MX warheads into orbit after a
Soviet launching. Whatever the domestically acceptable solution, militarily the system seems redundant.

Such plans for war arise from the compromise between geopolitical perceptions and logistic realities. If the deployment necessary to
achieve a political end is logistically unfeasible, then to attempt action
is ridiculous. The US Operation Blue seeking to free the hostages in
Teheran in 1980 seemed to verge on this. On the other hand, if the
most calculated, efficient, and flexible operational system was devised
in response to a flawed judgement of global politics, it goes for naught,
or, because of its potential for accidental triggering of nuclear bombardment, for much less. If the view of the Soviet Union as an aggressively

expansive empire with Western Europe as a primary target is untrue, then all the mathematical precision and technical brilliance of the nuclear deterrent constitutes merely a weapon for suicide, creating short-lived wealth and prestige for particular constituencies and interests as a side effect.

Logistics and Strategy

A major strategic instrument is being forged now which admirably illustrates the interplay of the geographical perception of the need for force and the costly minutiae of overcoming space and distance to deliver it. This is the RDF (Rapid Deployment Force) which is being put together for employment in southwest Asia.

The genesis of this force lies in the notion of an arc of crisis centred on the Persian Gulf with soft spots where Russian aggression could capture the oil fields which supply most of Japan and Western Europe's energy needs and 15 per cent of US needs. More immediately, the RDF was the Carter administration's reaction to the Islamic revolution in Iran and the Soviet invasion of Afghanistan. The Iran-Iraq war reinforced the fears of instability and loss of control in the Gulf. The RDF was to be the stick to enforce Carter's Gulf doctrine. In 1980 he announced to the world that the US would defend its vital interests in the Persian Gulf region against any outside effort to gain control and would repel such an assault 'by any means necessary, including military force'. Reagan has been less diplomatic and warned Russia that moving in on the Gulf, would 'be risking confrontation with the United States'. To fulfill this threat of action in the possible settings of Pakistan, Iran, Iraq, Saudi Arabia or the Strait of Hormuz, US firepower would have to be brought to bear from 7,000 miles away. The Soviets would be operating only hundreds of miles from their borders.

The Pentagon's analysts evidently foresaw that the bluff of strategic or tactical nuclear attack to defend someone else's source of oil, never mind the Iran of the mullahs, would be called. The horizontal escalation ploy, threatening to attack Cuba or the South Yemen if Russia sweeps into the oil fields, does not seem a trade which would impress the Kremlin. The only proposition thought likely to back up the declaration effectively was to counter men with men. This implied devising a means of closing fast with sufficient force to stop a Russian penetration or any insurgent take-over they fomented. Out of this was born the RDF. Critics castigated it at first as another example of the

Carter method of problem solving: 'throw a headquarters at it'. General Paul X. Kelley was given a staff of 260 in Florida and promised that if a flare up occurred in the Gulf he could draw on the 82nd Airborne Division and the 101st Air Assault Division for men. In its early months the RDF remained a paper tiger, 'neither rapid, nor deployable nor a force' in the words of James Schlesinger.

The arguments for and against the RDF turn on a question of logistics. There are those who see the Force as a mere political gesture, duplicating the marines' traditional role of 'kicking down the door' from a shipboard base. The marines' supporters proposed that they should retain this role, making amphibious assaults to provide a beachhead for troops airlifted from the USA. This strategy seems to be losing out to the case for a heavier forward contingent, with army units possibly based in very close proximity to potential flash points. The analysis involved has not been made public and interservice rivalry comes into play in the debate, but government actions appear to be reinforcing the forward base solution.

If Carter's actions were a reaction to events in Iran, the Reagan administration seems to have accepted the logic of a special formation for this theatre. Carter committed $10,000m to build up the force, hastily cobbling together some existing components. Two aircraft carrier groups in the Indian Ocean were supplemented by seven modified freighters carrying support for 12,000 marines and twelve fighter squadrons, including water, fuel, 50 tanks, 95 landing craft and 600 trucks. 1,800 marines were put aboard the Indian Ocean fleet during the Iran hostage crisis. Eight fast container ships and a Seabee barge carrier were bought to move a mechanised division to the Gulf in two to three weeks. Bases, air strips and ports of call were secured in the region. The UK loaned Diego Garcia and the atoll's runways and docks were enlarged to take B-52s and bigger troop ships. The US has permission to use Mombasa and Perth as liberty ports. There are airfields in Nairobi, Cairo, Guam and northwest Australia. Closer in the US has ports and airstrips at Ras Banas in Egypt, Berbera and Mogadishu in Somalia and Seeb and Muscat in Oman. Airstrips are available at Thamarit, Salalah, Masirah and Qus in Oman, close to the Strait of Hormuz. A satellite was launched to monitor movements in the Indian Ocean more closely. Plans were laid for a dozen pre-positioned ships to support a marine division (13,000 men) which would be permanently in place with the fleet in the vicinity of the Gulf by 1987. The planning objective was to get 300,000 men to the scene of any action as fast as possible. Plans to expand and improve the airlift capacity of the US were in train before the RDF was conceived. Two hundred CX transports which can land on short airstrips were on order, existing equipment was due for refurbishment and there was a plan for the use of civil aircraft.

The Reagan Secretary of Defence, Weinberger, confirmed the committment to the RDF and veered towards the overseas base strategy. In April 1981, two airborne divisions, a parachute division and a helicopter borne division, the most mobile elements of the American army, were consigned to the force. Future reinforcement is planned. In organisational terms, the force is being established as an independent command covering the Gulf and Indian Ocean. Formerly, it fell under the Army Readiness Command with its geographical targets divided between the European command and the Pacific fleet. A headquarters close to the potential fighting ground was being sought.

By mid 1981, the US could fly 800 paratroopers of the 82nd Airborne into any flash point within two days and a further 3,000 within two more days. Twelve thousand marines can be delivered from the Mediterranean and Diego Garcia within two weeks. These could be joined by a mechanised division of 15,000 men within a month and an armoured division within two months. With the commissioning of the prepositioned fleet by 1987, 13,000 men could be flown in within a week to man the 300 tanks and other hardware from the ships. The mechanised units could be landed within two weeks and the armour in a month.

In the interim, before the RDF is fully mustered, the Pentagon is depending on a tripwire strategy. It is hoped that dropping even a limited US contingent with plentiful air support to stand astride the rugged paths across the ridges and plateaux of south Asia will cause any Soviet advance to hesitate long enough to bring the political process into operation. It is noteworthy that elements of the 82nd Airborne were assigned to the Sinai peace keeping force in March 1982, establishing a command presence in the region and a desert training ground, at least.

The argument in favour of locating troops and supplies in proximity to the Gulf is to a large extent a matter of showing the flag, providing a presence to express US determination concretely. Initially directed to put Russia off an Iranian adventure, the function of the RDF has been expanded in geographical scope and required flexibility. Its targets are now not only any Russian probe across the Iranian plateau, the valley of the Tigris and Euphrates or Baluchistan, but also possible airborne assaults on the Strait of Hormuz or defending oil wells against any insurgent effort in Saudi Arabia.

The opposition to the RDF revolves firstly about its purpose and secondly around the feasibility of the deployment proposed. There are those who hold that it would be better to use the lighter, more mobile marines in their traditional role of dealing with small wars and guerrillas. There is serious doubt that the Russians are ambitious to spread themselves more thinly and risk any direct confrontation with Western Europeans and Americans over this resource area where they are least likely to give ground. If the Soviets intended anything the build up should be easily detected at an early stage, giving a long lead time for a counter. The national armies of the region are likely to offer a fight against any invasions. If there is truth in these propositions, then the RDF as planned looks over-elaborate and aggressively provocative.

Even if the strategic efficacy of the purpose is admitted, there still remains the question of feasibility. The limited manpower of the US is made less effective by having a large contingent based in the Gulf. Speedy air- and sea-lifting can bring them from America to Europe, the Gulf or any place required. There is a larger degree of inflexibility involved in the committment of the RDF to the Gulf.

There is, however, a counter argument in terms of the geography of the situation now which makes the case that the lead time is insufficiently long for the US to respond adequately to a massing of Soviet armoured divisions in Armenia or Turkmenistan. It has been customary for NATO to match Warsaw Pact May manoeuvres with a simultaneous show in Europe, calling up reserves and flying troops in from the USA. This is designed to deny the Soviets the temptation to prepare for an actual attack under the guise of annual exercises. If the USSR publically announced it was exercising in Armenia one year, with the secret intent of mobilising for a drive to the south, NATO would be in an awkward predicament as things are. To fly troops into an ostensibly neutral country to counter the USSR conducting manoeuvres on its own soil would appear provocative and might provide the Soviets with a plausible excuse to invade the Middle East. The longer reach provided by the local presence of the RDF may avoid the US being drawn into too provocative a stance in this arena. In addition, as things stand, the air- and sea-lift capacity needed to respond as it does in Europe, in the US civil airlines and US owned ships flying flags of convenience, could not be mobilised for this purpose for political reasons.

The extent of the theatre itself, however, and the uncertainty about where the opposition might strike means that the distance from any possible base to the fighting may be great. One view is that it would be better to avoid concentrating supplies in a base in favour of having them on board ship ready for direct delivery. The variety of physical and cultural landscapes encountered works against the notion that a local base will prepare troops for local conditions. Sandy desert on the Gulf is a far call from the rugged northern uplands or the densely settled banks of the two rivers.

As yet the base for the force has not been acquired. There is a reluctance among rulers of the region to have American soldiery in large numbers stationed in the midst of Islam. A quarantined cantonment, such as the oil company compound at Ras Tanurah, might be negotiated with the Saudis or Omanis. The prospect of a change of regime, bringing in an unsympathetic government which would renege on the agreement and throw the RDF out, with a subsequent loss of face not to mention costly emplacements, further militates against this solution. The debate continues, setting the political objectives of perceptions of the world against the realities of the friction offered by the earth's surface to military manoeuvring.

As we write, the British expedition to South Georgia and the Falklands following Argentina's invasion in April 1982, is demonstrating the difficulties of 7,000 mile supply lines and how they can be overcome at a cost if the political will is strong enough. However, in longer run geopolitical terms, the British Foreign Office has been seeking to contrive with Argentina something like the Hong Kong lease arrangement. Clearly, military responsibilities this far-flung are costly. A contrivance is necessary which will simultaneously salve Argentinian pride and protect the wishes of the small number of Falkland islanders, thus avoiding the prospect of such logistically expensive operations in the future.

Readings

The military objectives and interplay of mathematics, economics and electronics in the foundations of operations research can be found in:
G.B. Dantzig *Linear Programming and Extensions* (Princeton University Press, Princeton, 1963), ch. 2

The nuclear logistic examples are treated fully in:
E.S. Quade (ed.) *Analysis for Military Decisions* (North-Holland, Amsterdam, 1970), ch. 3
H. Scoville *MX: Prescription for Disaster* (MIT Press, Cambridge, Mass., 1981)

Discussion of the RDF is from news sources, especially:
The Economist, 6th June 1981, 'Defending the Gulf: A Survey'

4 TACTICAL PLOYS

It is because of disposition that a victorious general is able to make his people fight with the effect of pent-up waters

Sun Tzu, (chapter 4, verse 20).

Tactics and Strategy

For the sake of exposition a simple distinction can be drawn between tactics and strategy. Strategy is choosing to fight at the right places at the right time, while tactics is making the best use of the force available in any given geographical setting to complete a task determined by strategic objectives. Tactics is about arranging men and weapons advantageously on a battlefield. Strategy is about selecting the sequence of battlefields. To succeed, both tasks must be done well. Either the wrong deployment on the right battlefield or good tactics on the wrong battlefield may lose the war. Obviously there is an interaction between the two levels of resolution. Battlefields should be chosen with the capabilities of weapons and manpower in mind. Tactics may be modified to achieve a strategic end. For example, a retreat where immediate tactical victory was possible may be called for by the larger scheme of things.

The difference between tactics and strategy has been confused by the rapid changes in the range and velocity of warfare of the last two centuries. Rifled weapons extended the distance at which small arms fire was exchanged from one hundred yards to three hundred and placed a premium on the skilled use of cover and camouflage. The range of artillery has leaped from yards to miles to continents with ballistic missiles. From the age-old three miles an hour of pedestrian infantry, the speed of manouvre has risen to the thirty miles an hour or more of tanks, troop carriers and trucks, and taken on a third dimension with air support and helicopters. Napoleon recognised the change of the appropriate geographic scope which resulted from the greater velocity of his divisions by adopting Guibert's term 'grand tactics' to describe his plans. The new extent of the battlefield brought tactical decisions outside the range of visual contact with the enemy. This introduced a greater potential for the employment of deception, the principle weapon of strategy. Tactics and strategy could no longer be distinguished in terms of whether eyeball contact was possible. The continuum of decision between what can be done on the fighting ground and the broader sweep of move and counter-move in the prosecution of a war becomes denser. However, the event that is a battle and the duration of a war remain distinct and this demarcates tactics from strategy. The first is concerned with the management of a battle, the second with the orchestration of battles in a war or campaign. This chapter concentrates on the deployment of men and firepower on a battlefield making the best use of the ground, leaving the broader issues to be dealt with under the heading of campaign strategy.

Offence and Defence

There are two fundamental tactical stances, the aggressive and the passive. The choice of one or another should be a matter of strategy. Whether to attack or defend, should be judged in terms of the time, space and force available to and confronting the commander. Great commanders, like Jan Žižka with his Hussite wagon-forts, combined strategic attack with the tactical defensive. They used mobility to manipulate the enemy into a position where he had to choose between attack and retreat. This gives them the advantage of the lie of the land, which usually accrues to the defender. The attacker is at the disadvantage of making a strong commitment to using force in a certain direction. Such a thrust can get out of control, developing a momentum of its own. Having carried the war to the enemy strategically, it has often paid to take the tactical defensive and let them make the first move in battle. The great tactical talent lies in turning from the defence to attack when the time is right. The defender's advantage of the theatre of war will only pay off if he can turn on the aggressor when his weakness shows. The precarious balance between the attacker's advantage of surprise and the potential danger of detectable overcommitment is the essence of most contests. In football, fencing or judo, whether you attack or defend should logically be a matter of what your strengths are compared to those of your opponent. You go in first if you think you have enough muscle to overpower him and enough quickness to recover in time if you miss, to meet his counterstroke. You hold off if you think you can beat him with a speedier riposte.

Historically the choice has been a matter of temperament and tradition. Tactics was a repetition of what proved successful in the past. The conventional wisdom that attack is the best form of defence and the spirit of *élan* and the *arme blanche* has led generations of soldiers to death in the face of new weapons, whilst elsewhere a tradition of passive, fortified defence has had similar effects from a different starting point. The Clausewitzian fervour for cold steel and concentration ignored the lessons of the defensive power of entrenched riflemen of the American Civil War to come to grief on the fields of Flanders. The Maginot Line mentality which then mastered the French military, brought defeat in an era of tanks and planes. Tactical success often comes from a collective temperament which is in tune with the weapons of the time and a tradition emerges to repeat what proved successful in the past. A temperament is often cultivated by the established order to sustain a tradition in tactics. Hunting to hounds in Britain was a means of fostering the cavalry spirit as well as the requisite skill. The temperament that prevails in particular military establishments can be ascribed to geographical and cultural circumstances. The disparity between the Royal Navy's inclination to dish it out and the British Army's tendency to stand and take it, can be construed as a result of Britain's insularity. The navy, in the British element, aggressively sought out and destroyed the enemy in the Nelson tradition. By contrast, the army, separated from its home base by the sea, was inclined to caution and the defensive. The tone was set by Wellington, choosing and holding his ground while the enemy wasted himself in attack.

A rational choice of whether to attack or defend would take into account the course of events, the geography and the forces available to and confronting the commander. To be predisposed to one or the other is to reveal your hand which will only bring victory if you have the strongest cards, unless the enemy is stupid. A known predisposition leaves whatever weaknesses there are open for exploitation.

Having decided whether to attack or defend, tactics becomes a matter of moving over and occupying the terrain so as to maximise the effectiveness of your force. Defence can obviously be static or mobile. The static defence seeks to present an impenetrable perimeter for the enemy to break against, while defending its line of retreat and supply. The defender digs in and commands the ground, creating a barrier which will cause the maximum damage to the aggressor and prevent any breakthrough or envelopment. A mobile defence seeks to suck the aggressor in to extend his lines and break up the concentration of his assault. This maximises the length of his front, spreading his force as thinly as possible, presenting the prospect of ripping out new flanks to chew upon.

Attack implies mobility and has the objective of penetration, rolling up the defenders' perimeter and ripping up his interior lines. By manoeuvre it is hoped to confuse the defence as to the direction to point in, so as to lay open any weak joints. The basic objective of attack is to concentrate a sufficiently superior force at some point or points on the defensive disposition so as to overwhelm it a bit at a time. There have been many variations on these basic themes and many engagements have involved each side in attack and defence in turn. Tactics is the art of positioning and moving forces in this setting so as to achieve a strategically determined goal given the weapons and terrain at hand.

Formations

The geographical arrangement of forces on a battlefield can be described in terms of three fundamental formations. These are the line, the column and the square. Much of tactics is concerned with which of these to adopt in a particular setting, when and how to change their bearing or when and how to change from one formation to another. With the increased range and power of weapons and communications, with air support and with the greater speed of manoeuvre, the density of these formations has decreased. Their geographical extent has increased and their outlines have blurred since the days of pikes or muskets. They do, however, still provide a useful set of general configurations with which to conceive the deployment of basic tactical elements.

The line confronts the enemy with the greatest length of cutting edge and is the most difficult deployment to outflank. In defensive form it constrains and contains the enemy and can be enhanced with fortifications and obstacles. The offensive line brings as much of an army's firepower to bear as possible in an advance. To avoid being overlapped the flank of a line can be anchored on an obstacle to movement. Hills, rivers, the coast, or cities have been used as geographical auxilliaries to a line. An escarpment rising above the flood plain and the bank of a river have been used time and again as end stops for a line drawn up across the line of passage in a valley. The line does have the disadvantage that it lacks depth and, thus, strength. It is also lacking in mobility. Keeping dressing so as to avoid opening gaps is difficult when men and vehicles are strung out across country. To travel off road is difficult in most varieties of terrain. The defensive line resting a flank on a geographical feature is rendered less mobile by the need to keep contact with its end stop. With the introduction of the rifle and rifled artillery the close order of the past gave way to thin skirmish lines. The line became a cloud taking advantage of cover and camouflage. The offensive line began to take the form of a racing flood, infiltrating through defences at weak points so as to attack from the rear.

The column sacrifices coverage for mobility and is the most vulnerable to a flanking attack. This is the formation of an army on the march. It is in effect the dual of line formed by turning its elements in the direction of a route. This duality takes the form of a narrow front and long exposed flanks. Building on the innovations of Carnot, Napoleon's early success was won by forging the revolutionary horde into loose columns headed by swarms of skirmishers. These flowed forward rapidly, seeking to concentrate force at the enemy's weak spots and confuse him with speed. As Napoleon saw it, 'The strength of an army, like that of a moving body in mechanics is expressed by the mass multiplied by the velocity'. Velocity was achieved with columns. Sherman demonstrated the value of columnar mobility in the American Civil War, which lesson was put into practice in the tactics of the blitzkrieg in the 1940s with a spearhead of tanks before a shaft of infantry.

The square is formed by joining the flanks of the line so that it faces all directions. This is the least mobile. In the coming and going of tactical fashion between dispersion and cohesion through history, the square stands at the cohesive limit of the spectrum. This most defensive stance guards against the flank attack. In the extreme its mobility becomes zero where it takes on the built form of fortifications. It has proven an effective counter to an enemy with larger numbers or greater mobility. It reduces the destruction and panic created by attack from the side or rear. The encircled laager of wagons has been the traditional defence of migrant tribesmen. If its movement can be coordinated so as to preserve its defensive integrity it becomes a powerful weapon of war. In effect a cross between the square and a column combines mobility with invulnerability. Richard the Lion-heart achieved this in his march from Acre to Jaffa in 1191, using the sea to protect one flank and a protective layer of infantry and crossbowmen on the landward approaches to shatter Saladin's attack. In the 1420s Žižka's wagon-forts provided mobile squares from which to smite the Teutonic Knights. In the winter of 1950 a US Marine division broke out from Chosin, fighting a thirteen day running battle against overwhelming odds formed in a walking square moving a mile in four hours. The static form of the square has also been successful in the right time and place. Peter the Great's redoubts raked the Swedish columns in Poltava in 1709 and brought Charles XII to grief. Wellington's squares of British infantry destroyed Ney's cavalry at Waterloo. The Wehrmacht's 'hedgehog' squares on the plains of Russia saved it from destruction in the winter of 1941-42.

In the 1980s, the increased range and lethality of conventional weapons and the threat of nuclear weapons has tended to make modern defensive tactics various combinations of small squares. Company or battalion size units occupy 'battle' or 'blocking' positions far enough apart so that not more than one defensive position can be destroyed by a nuclear weapon. These widely dispersed defensive positions are organised for 'all-round defence' (the objective of the square) and are supposed to cover the wide gaps between positions with machine gun, tank gun, and anti-tank missile fire. Troops lacking mobility equal to the enemy's (such as light infantry) would ideally occupy 'tank-proof' terrain such as villages or forests from which they would engage by-passing mechanised attackers with long-range anti-tank fire. A defensive line will consist of a string of these dispersed company or battalion positions. A mobile defence will consist of a 'checker board' of these positions which would be occupied in succession by 'blocking' units attempting to channel the attack into a pre-selected nuclear strike or counterattack 'killing ground'.

Aggressive Tactics

For the side which takes the tactical initiative there are three elementary moves to employ: the frontal assault, the outflanking attack and the wedge driven to split the enemy's formation. Combinations of these, feints and the exploitation of the lie of the land and the enemy's disposition and reactions gives rise to the variety of tactics.

The frontal assault is the crudest approach. Colliding belly to belly with the opposition in the manner of a sumo wrestler does not show a great deal of military flair or imagination. If you have a superiority of numbers, weapons, technique, discipline or zeal, however, it may pay to bring it to bear in the most direct fashion and settle the matter conclusively. This was the spirit which led the Roman legions to scorn subterfuge and employ their flexibility and discipline straight ahead. When their time came, the Arabs of the Islamic explosion swept over the opposition in wave-like attacks. In the 1660s, Martin Tromp used the superior seamanship and firepower of the Dutch fleet in the naval equivalent of the frontal assault, bringing his guns to bear by swinging broadside to the enemy to deliver his artillery assault. At Lodi, Napoleon created the legend of French invincibility by the deliberate, bloody, frontal assault of grenadiers against Austrian guns.

If he did not prescribe the frontal assault in detail, von Clausewitz's writing was imbued with the spirit of directness, of seeking a decisive outcome in battle. As he put it 'the battle . . . the bloody solution of the crisis, the effort for the destruction of the enemy's forces, is the first born son of war'. His writings were philosophical rather than tactical but the spiritual obsession with cold steel and blood they engendered in the Prussian military and their imitators led to an inclination to full-frontal confrontation. With a near equivalence of forces and weapons the artless adherence to direct and total engagement can become a formula for mutual and wasteful destruction. In tactical terms, the current deployment of nuclear weapons by the USA and the USSR has all the appearance of preparation for a head-on clash, but with the globe as the battlefield and their cities as the front line.

The flank attack seeks to strike the enemy in the side or back. Its most primitive form is the ambush which takes advantage of the terrain, originally the availability of bushes, to deliver a surprise blow to an enemy. Hannibal smashed the legions of Flaminius in 218 BC coming out of hiding in the woods to catch them in column of route against the shores of Lake Trasimene. Ambush is the classical tactic of guerrillas against the superior arms of invaders or the established order, hitting the exposed flanks and rear of forces on the move. Napoleon used the superior speed of his divisions to sweep to the enemy's rear, cutting off his line of retreat and supply. In naval tactics the manoeuvre of crossing the T, sending a broadside line across in front of a column to bring greater firepower to bear is a pure form of flanking attack. This was the basis of Togo's defeat of the Russian fleet at Tsushima in 1905. The tactics of the blitzkrieg and more recent mobile warfare constantly seek to outflank the enemy and get at his rear, now with the added dimension of aircraft providing for vertical flanking movements striking from above.

The flank attack has most usually been combined with a frontal attack. If the enemy is not caught by surprise on the move but is in battle order or is not greatly outpaced, then the approach from the side or rear must be in concert with a frontal engagement to hold his attention. The frontal assault acts as an anvil against which the hammer of the flanking troops can strike the enemy from a direction in which they are vulnerable. The flanking movement developed naturally from the frontal assault where superior numbers enabled one side to curl around the end or ends of the other's line and roll it up. The first recorded cases where the flanking was contrived was in the Theban victory over Sparta at Leuctra in 371 BC. Epaminandos used what has come to be called the refused wing tactic to capitalise on the known tendency for the right wing of a line to advance more rapidly, as each hoplite sought the cover of the shield of the man on his right. He deliberately advanced the Thebans obliquely with the left wing leading. This in effect refused to engage the Spartans' trailing left wing and concentrated the attack on their advanced right wing, outflanking and rolling it up.

Envelopment is the tactic of overlapping both enemy flanks with a pincer-like movement. This is made easier if the enemy can be induced to break or bend the ridigity of his formation and step into the jaws of the pincers. This can be achieved in a number of ways. Chance gave Miltiades victory in this fashion at Marathon as his centre retreated and drew the Persians in between the horns of his advancing wings. At Cannae, Hannibal presented the Romans with a bowed out centre which he then drew back to bend inwards in an orderly withdrawal. The legions followed into the waiting trap of the outflanking Carthaginian wings. This tactic of the refused centre, sucking the enemy in, was turned on Hannibal by Scipio at Ilipa. In this case the two Roman wings moved forward more rapidly than the centre on a static Carthaginian line. This thrusting envelopment was the ideal of von Moltke's Prussian war machine, an advance which fastened on the enemy's centre and swept around their flanks to smite their rear. The use of a holding frontal attack combined with a double envelopment is graphically portrayed by the Zulu division of their tribal troops into 'chest' and 'arms'. The arms were supposed to manoeuvre around and behind the enemy and crush them against the centre or chest.

With the increased range and speed of engagement and reduced density of formations of the last century, they have become more permeable, flexible and widespread. The geometrical formalities of the past have given way to the swirling, opportunist movements of encirclement and infiltration, like Rommel's cauldron tactics in the Western Desert. The objective, however, remained the same: to get past the foe and menace his rear.

The third aggressive tactical ploy is penetration by means of a wedge driven into the opposing line. This can be considered another special case of the flank attack. The opponent's formation is wedged open to create two new flanks to roll up his line from the side and back. The classical example of this manoeuvre was at Arbela where Alexander's cavalry breached the Persian line to panic and destroy their superior numbers. Nelson's masterly victory at Trafalgar drove two columns into Villeneuve's line, breaking it up into three isolated pieces, which superior British gunnery then went to work on. Napoleon employed a wedge at Borodino, albeit in a heavy-handed, brutal fashion. By this stage he had regressed by shifting the emphasis in his tactical equation from velocity to mass. The French army was put into one great column to smash into the centre of the Russian line. Over the last several decades mechanised columns have churned in a variety of penetrating wedges along with pincers and flanking movements seeking any weakness in the elongated, permeable fronts of mobile war. The size and dispersion of modern armies has made the penetration the standard offensive tactic. In World Wars I and II, the size and weapons range of the engaged armies rarely provided open flanks. A NATO-Warsaw Pact or China-Soviet Union conflict would also most likely fill the respective war zones with continuous fronts.

Defensive Tactics

In dealing with line and square formations we have in essence considered the principle tactics of defensive warfare. There is, of course, a distinction to be drawn between static and mobile defence. Whichever of these we consider, however, it is plain that the defence has a better chance to pick its fighting ground and, thus, to use the lie of the land to its favour. The terrain can be turned to defensive advantage providing cover and friction to brake the aggressor's mobility. Wellington, for example, always carefully selected the position he would let the enemy come on to, stationing his troops behind the brow of a hill protected against artillery until they needed to engage. Slopes and rivers or marshes can be used to slow the enemy's advance and increase his vulnerability to missiles. The introduction of rifled guns and then tanks and aircraft to warfare have placed a premium on the tactical skill of using the landscape as cover, concealment and an obstacle.

The static defence can enhance the advantages of the landscape or overcome its lack of cover and friction by making fortifications against missiles and artificial obstacles to movement. Such works can in general be linear or square, seeking either to turn the enemy tide on a long front, or break up its momentum by controlling crucial places in the land. In the 17th century Vauban raised the engineering of fortifications, and their destruction, to a fine art in the competition for mastery of the river and canal towns of the Low Countries. After 1850, railway networks became the logistic sinews of nations and the object of defensive control and offensive destruction. The Wehrmacht's hedgehogs in Russia, for example, were built around railway stations. The coming of the railways also ushered in the era of massive armies with powerful long range weapons. The tactical implications of these changes could be observed in the American Civil War, where the first command of a million men fell to Grant. Rifles and artillery gave the advantage to the defence and all the components of the entrenched stand-off were present in the lines outside St Petersburg in the valley of the Appomattox in 1864. The lessons to be learnt were ignored by Europeans and this was dearly paid for in 1914-18. The tactical deadlock was broken in April 1917 at Cambrai where the tank made its debut, shifting the balance in favour of offence rather than static defence. The offensive three dimensional tactics of the blitzkrieg operating at 30 mph on the ground with tanks and trucks, combined with air support and airborne troops could easily overwhelm 'hard thin skin' defences like the Maginot Line. The more permeable but deeper Siegfried Line, with its 'Milky Way' of pillboxes and tank traps was not designed as a last ditch but as an expendable friction zone to slow and bleed an armoured advance. Now with planes, helicopters and guided missiles, elaborate fortifications would seem to be of little use since they can be hopped over if not curled around. The last bastion of defensive fortification may be the hardened missile silo and even these are possibly vulnerable to penetration and disruption by the electro-magnetic pulse of nuclear explosions.

What deep offensive penetration called for by way of defensive response was mobility and flexibility. There were many historical precedents pointing the way. The delaying tactics of Fabius were taught with the classics. The 500 years of Byzantine success in letting the foe come forward and then slashing him to pieces with brilliantly deceitful manoeuvre was less well-known but, nevertheless, personified in the legendary Belisarius and available for study in the *Strategicon* of Maurice and the *Tactica* of Leo the Wise. The methods of insurgents and partisans had been used to good effect time after time. Faced with the tidal wave of Barbarossa the Russians' memory of 1812 was quite fresh and they responded with an elastic defence in depth. Islands of resistance were left behind to sap the enemy's power by attrition. The earth between these pockets was scorched of sustenance for the enemy. Cossacks gave ground before the onslaught at a rate of 18 miles a day, but maintained their attack on the German front with mobile firepower.

Giving ground in exchange for time, the Wehrmacht was drawn out, cut up and driven back.

Tactics Today

The porous fluidity and speed of mechanised warfare with radio communications and airborne firepower has extended the scope of a continuous engagement over time and space. Set piece battles are replaced by sprawling conflicts with sporadic fire fights. The manipulation of formations in broad sweeps to outmanoeuvre the enemy has become a matter of strategy rather than tactics. The scope of battle has gone far beyond the ken of one man's unaided perception of what was going on over a stretch of country. The geometrical regularity sought in formations in the past, where the simplicity of drilled movements in response to simple signals was a means of achieving control and concerted effort, is no longer necessary. Tactics, the art of arranging men in the field of battle within the perceptual grasp of a single commander, then dissolves to the detail of deploying men and weapons for a given task so as to take advantage of opportunities presented by the theatre of war and the actions of the enemy within your local scope. The command of armies becomes a matter of strategy. What is left to tactics is for the most part the indoctrination of soldiers with common sense ways of fighting, using their weapons and vehicles to best advantage while avoiding the enemy's deadly attention.

Given the task of stopping the enemy advancing over a piece of territory or taking it off of him, warlike manoeuvres can be likened to water flowing over a surface of military opportunity. Units gush along open channels, flow around resistant obstacles and eddy into convenient, sheltered pools, seeking the line of least resistance for an advance or to soak the land so as to mire the enemy's advance. Either advance or defence involves a thorough appreciation of the terrain if it is to be done effectively. As a first step it is necessary to identify geographical targets and places to avoid. Desirable locations are either places you want or want to deny to the enemy. These can be either natural or man-made geographical features which afford their holders a military advantage. Places which are exposed to enemy surveillance and fire are obviously to be shunned if possible. Desirability is mostly a matter of protected places from which to command the theatre. Good ground provides the means of keeping an eye on the movements of the enemy and inflicting damage on him with your weapons, or else, from the opposite perspective, it gives shelter from enemy fire and a screen against enemy observation. Having determined where to get to, or where the enemy may wish to get to, the next step is finding how to get there with the minimum effort and loss of life, or how the opposition might approach target locations. This involves gauging how easy it is to traverse the surface. The routes available and the natural or artificial obstacles which prohibit or inhibit movement of men and machines must be surveyed. This involves not only the quality of the roads but also the going off the roads and how this changes with weather. Lines of movement to and from the targets and the speed with which they can be travelled need to be established. The analysis of the landscape in terms of military attractiveness or its inverse, dividing it into places you want to possess or those you wish to avoid, and the accessibility offered by routes and the lie of the land, provide the basic information for manoeuvering and deploying men and weapons. Beyond this the details of paths and placements chosen are a matter of fitting the capabilities of arms and vehicles to the terrain and second guessing the opposition.

The most readily available example of the detailed procedures are those of the US Army as given in its field manuals. The political premise which underlies these documents is that the most likely call for action will be a defensive war against the Soviet Union in Europe. The strategic assumption is that NATO will face a superiority of men and armour and will fight in an orderly, gradual withdrawal, winning time for a negotiated settlement. The tactics then will be those of an active defence, retreating in order, delaying the Warsaw Pact forces and causing them as much grief as possible. The geographical component of US Army tactical analysis is summarised by the acronym OCOKA. The five aspects of the terrain captured in this are observation, cover, obstacles, key terrain and avenues of approach. The reference to observation signals the soldier to consider the area he is operating in from the point of view of the opportunities it presents for surveillance and fields of fire. A field of fire is the ground that a weapon can cover from a particular location. The dual of this involves seeking cover, that is, protection from enemy fire and concealment from his observation. Obstacles should be scouted out either so as to avoid them or to employ them against an advancing enemy. Key terrain refers to commanding features in the landscape, and avenues of approach are the routes to these or to operational objectives.

Rules and advice to encourage an awareness of these terrain characteristics are incorporated in the more specific doctrine directed at the three levels of tactical command for tank and mechanised infantry combat. The elementary level is that of the platoon or squad where the emphasis is on the actions of the individual soldier. The next level is that of the company where the commander fights the battle on the basis of what he can see. The fundamental tactical level is the battalion command from whence the battle is controlled, fitting component units to the terrain with what is in essence the exercise of map reading skill.

The chief objective of platoon level indoctrination is to get soldiers to use cover. They are advised to seek frontal cover and to avoid land-marks, walking on the skyline and kicking up dust. They are encouraged to move quickly in the open, having marked cover they can run for if attacked, and not to mask their covering fire. Particular attention is paid to the limits imposed on the performance of weapons by terrain. One notable case is the TOW (tube launched, optically guided, wire controlled) heavy anti-tank missile. Given that the missile takes five seconds to travel 1,000 yards and that it takes some time for the operator to latch onto a target and fire, for an engagement at the range of a mile, a tank must remain exposed for 100 yards or so to be vulnerable. If it can scurry behind a hill, building or clump of trees it is safe. The weapon is best brought into play where the terrain allows of constant visual contact. Woods, hills, towns or dust can nullify its value. The weapon is best used in flanking fire from a hidden position at a range of 2,000 to 3,000 yards. This stand-off is the distance at which the TOW can engage a tank yet remain out of its range. Such detail provides clear tactical guidelines.

The company commander's main concern is with enemy avenues of approach and the whereabouts of supporting forces. In assigning men and weapons to battle positions, cover and commanding locations are sought for long range anti-tank fire and shorter range crossing and flank fire, disposed so that obstacles impede the enemy's movement through fields of fire. It is possible to be quite specific about what will slow up a tank. Tanks have trouble with slopes over 30°; sharp rises over 5 foot; gullies or ditches over 15 foot wide; rivers or canals over 150 yards wide and 5 foot deep; bogs deeper than 3 foot; forest with trees 8 inches in diameter; trees 4 inches thick on a 10° slope; tree stumps 18 inches high; snow 3 foot deep or a landscape of buildings.

The battalion task force commander must direct matters beyond the range of vision and relies heavily on maps. Despite its penchant for writing everything down, the US Army has not prescribed a standard, systematic method for terrain analysis at this level, although there are some suggested rubrics in circulation. These involve a variable sequence of identifying obstacles to armour and the paths from where the enemy is now round the obstacles to his likely objectives. The alternative paths are ranked according to the likelihood of their use, and forces are allocated between these according to the potential danger. For each avenue, key terrain, observation points, fields of fire and cover and concealment are evaluated to deploy the firepower allocated to the best advantage.

Competing tactical theories were put into practice for all to see in the Six-day War in 1967. The Egyptians employed the standard Soviet 'sword and shield' defence. This involves placing a three layered shield across the enemy line of advance and holding a sword of tanks behind it in case they breach it. The first layer of the shield to confront the enemy is an artillery fire zone; the next is a minefield and, finally, a deep layer of infantry armed with mortars and anti-tank weapons. To the rear of this stands the artillery fire base and the mobile reserve of a tank force ready to meet the enemy armour wherever it breaks through the shield. Against this the Israeli's brought their 'ugdah' formations, imbued with the spirit of aggressive fluidity. Tank columns, followed by mechanised infantry, flank the route of advance. Lagged behind the armour on the road comes the main supply column, with a shuttle service providing a constant supply of fuel and ammunition to the spearhead. The tank columns converge and drive a wedge through any enemy defensive line and roll on. The mounted infantry widen the breach and chase on after the tanks. It is left to foot soldiers to mop up the resistance and make it safe for the supply column to advance along the road. This arrangement suffered from a tendency for the spearhead to tear on leaving the infantry and supply column faced with an opposition that had regrouped after recovering from the armoured punch. What gave Israel's offence the edge in this war was the monopoly of air power they enjoyed after their surprise destruction of the Egyptian air force on 5th June. In the Yom Kippur War of 1973, the Egyptians had SAMs (surface to air missiles) which kept the Israeli air force at bay. It was only when they lured the Egyptian tanks out beyond the range of their SAM batteries and could employ their superior mobility and flexibility that the Israelis got the upper hand.

With this superficial treatment of tactical history and current tactical doctrine and practice, we are in a better position to appreciate the relationship of tactics to terrain dealt with in the next chapter and have a basis for examining strategic matters later.

Readings

The historical portion of this chapter is indebted to:
Lynn Montross *War Through the Ages* (Harper & Row, New York, 1960)

The US Army Field Manuals invoked here are listed in the readings for chapter 2.

Systematic schema for tactical terrain analysis are to be found in:
Lon E. Maggart 'An Analytic Approach to Terrain Analysis and Allocation of
Combat Power' *Military Review*, vol. 58, no. 4, 1978, pp. 34-45
Joseph C. Gross III and Harry B. Beam 'Terrain Analysis' *Armor*, vol. 88, no. 1,
1979, pp. 17-20.

Maggart's efforts found their way into official circulation for offensive tactics as:
US Army, *Lesson p. 312-22, Section II, Analyzing and Developing Avenues of
Approach for the Offense* (Fort Leavenworth, US Army Command and General
Staff College, 1978)

and for the defensive tactics as:
US Army *RB 100-9, A Systematic Process for Analyzing Avenues of Approach
and Allocating Combat Power for the Active Defense* (Fort Leavenworth, US
Army Command and General Staff College, 1978)

The Egyptian-Israeli wars are analysed in essays by H.P. Willmott in *War in
Peace* (Sir Robert Thompson ed.) referred to at the end of ch. 1

5 TACTICS AND TERRAIN

> Conformation of the ground is of the greatest assistance in battle
>
> Sun Tzu, (chapter 10, verse 17).

Tactical manuals and military histories frequently make casual use of various terms when describing the terrain on which a battle is being planned or has been fought. Prominent among these rarely defined terms are 'corridor', 'cross-compartments', 'close terrain', 'open terrain', and 'normal' versus 'extreme' combat environments. Another frequently used term, although one usually more carefully defined, is 'key terrain'.

In this chapter we will examine these concepts as illustrated by historical examples. We will point out the importance of the scale of operations and the nature of the combatants in the application of these concepts. Finally, we will examine how the nature of the tactical military geography of a region can change radically with changes in the nature of warfare.

Corridors and Cross-compartments

The term corridor implies terrain favouring the movement of military forces bounded on either side by unfavourable terrain. The term cross-compartments implies linear terrain features unfavourable to military movement that extend across the desired direction of movement. The same features may form either cross-compartments or corridors, depending upon the direction of intended operations.

In the US Civil War the north-south trending linear ridges and valleys of the folded Appalachians formed corridors during the Valley Campaigns in western Virginia, but formed cross-compartments in the campaigns in Tennessee and Georgia.

The attacking armies of Stonewall Jackson in the early part of the war and those under Phillip Sheridan in the latter stages of the conflict found the broad limestone valleys adequate for deploying armies of the size and nature of those employed. The north-south trend at these valleys facilitated both Jackson's offensives north and Sheridan's offensives south. Defending forces found no strong defensive barriers across these valleys and could easily be forced into battles of manoeuvre on the broad valley floor. Furthermore, a strong defending force in one valley could be flanked by the attacker side-stepping across a ridge to an adjacent, less well-defended valley. This tended to cause defending armies to dissipate their forces by trying to block a series of parallel valleys or corridors.

Very similar terrain in the southern Appalachians, however, formed cross-compartments. Here, Union forces under Rosecranes and Sherman advanced against Confederate armies led by Bragg and J.E. Johnson, respectively. The Federal armies were advancing from west to east, at right angles to the north-south trend of the mountains and valleys. The Southern armies simply occupied the railway pass through a ridge, forcing the Northern army to either launch a frontal assault against the strong ridge-top Confederate defences, or to abandon their logistical life-line of the railway and launch a long turning movement north or south to cross an undefended pass. If the Confederates discovered the Federal movement in time they could shift to defend the pass that the Northern troops were headed for. If they didn't detect the Federal movement in time, the Rebel forces simply fell back across the valley to the next railway pass or tunnel through a ridge. The Union armies, unwilling to operate for long periods far from their logistical life-line of the railway, were thus faced again with either storming or turning the Southern position by wide marches.

In contrast to the fast-paced battles of manoeuvre of the Valley Campaigns, war in the southern Appalachians was a series of bloody, inconclusive frontal assaults on strong positions, or lengthy and equally inconclusive turning movements. The armies and the terrain were similar, but in western Virginia the armies were moving north and south along the corridors of the broad valleys, while in Tennessee and Georgia they were moving west to east across the cross-compartments of the parallel ridges and valleys.

The eastward flowing rivers and creeks of northern Virginia were another series of topographic features that formed corridors and cross-compartments across strongly contested turf in the Civil War. Along the Washington to Richmond route the southward movements of the Army of the Potomac were contested by the Army of Northern Virginia, and battles were often fought where the Union troops attempted to cross a river or stream. The battles of Bull Run where the Confederates blocked the Union crossing of the creek by that name, and the bloody battle of Fredericksburg, where Lee held the bluffs of the Rappahannock against Burnside's river crossing, are noted examples. These drainage features formed a corridor, however, when McClellan attempted to reach Richmond by landing the Army of the Potomac at Yorktown and advancing west between the Rappahannock and the James. These rivers protected McClellan's flanks while he advanced on Richmond. This 1862 Union campaign was halted less than 10 miles from Richmond in the desperate battles of Fair Oaks and Mechanicsville.

Although corridors and cross-compartments are usually thought of in terms of land forms or drainage features, bands of vegetation may create the same phenomena. The German drive across the north European plain towards Moscow in operation Barbarossa during World War II encountered cross-compartments created by forests. Forested areas in this part of the USSR tend to be arranged in north-south bands along rivers and poorly drained areas. The higher, better drained rolling plains between these forest belts were usually open farmland. The Wehrmacht's eastward advance often took the form of a rapid dash across the open belts with mechanical and motorised forces followed by a slower, hard fought advance along narrow forest roads in the wooded belts. The Pripet Marshes, between the northern and southern wings of the Nazi advance, formed a forested corridor for Soviet cavalry units. Swept from the open plains by German mechanised forces, horse cavalry, the ancient weapon of the steppes, found a new role as a mobile striking force in roadless forests. Horse-mounted Soviet divisions could infiltrate from east to west through the forests of the Pripet and strike the Wehrmacht flanks and rear.

Close and Open Terrain

Close versus open terrain usually refers to engagements in built-up areas and forests (close terrain) as opposed to farmed or grass covered plains (open terrain). The latter circumstance is usually regarded as normal by modern armies while combat in cities or forests is special. The classification of terrain as close or open seems to be based upon the likely range of engagement of opposing forces. Combatants tend to fight each other at closer quarters in close than in open terrain. Along with range of engagement, casualty rates, rates of advance, and degree of control over subordinate units vary with the closeness of the terrain. Modern direct-fire weapons tend to have effective ranges far in excess of the range of engagement in cities and forests. Heavy firepower at close ranges usually results in higher casualty rates than when similar units engage at longer ranges in open terrain.

Rates of advance tend to be much slower in cities and forests than in open plains. This is especially true for modern mechanised forces. The Germans dashed across open spaces and slugged through forests in European Russia in World War II. On the western front, Patton toured France with an army as the armoured divisions of his Third Army raced across northern France after the breakout from Normandy. When logistics permitted the continuation of his advance through the broken and forested Lorraine, however, movement was slow and his corps commanders called for more infantry to assist in the close forest encounters. The defending German generals also requested more infantry for the jungle fighting of Lorraine.

Mechanised forces tend to by-pass urban areas whenever possible, for progress through cities is excrutiatingly slow. Paulus' Sixth Army raced across the steppes of the Ukraine in 1942 only to grind to a block by block pace in Stalingrad. Zhukov's Red Army marched from Warsaw to the Neisse-Oder in the month of January, 1945, against the disintegrating Wehrmacht. Yet the battle for Berlin itself took from 16 April to 8 May and cost the Soviets at least 300,000 casualties.

Most armies seek battles of decision in open terrain, where control of subordinate units and battles of manoeuvre are possible. Battles in close terrain, such as forests, tend to disintegrate into 'soldiers' battles' as higher headquarters lose control of the fight and small units decimate each other at close range. In the US Civil War many battles tended to become bloody soldiers' battles because of the second growth forest that divided units and led to close engagements on so many battlegrounds. One of the largest battles ever fought between Anglo-Saxon armies (over 100,000 men engaged) was between Rosecranes and Bragg, as their respective Union and Rebel forces groped for each other along West Chickamauga Creek in western Tennessee in September, 1863. Both sides lost about 40 per cent of their infantry strength in this bloody and confused engagement, and half the Union Army was driven from the field. Yet Bragg's victory proved indecisive as he could not control his forces enough to pursue and destroy Rosecranes' defeated Army of the Tennessee. Not only are line of sight and messenger communications disrupted in forests and urban areas, but the radio communications of modern armies is also much reduced in effectiveness. It is this difficulty in communicating and the difficulty for commanders to 'see' the battlefield that results in loss of control and the resulting soldiers' battles in close terrain.

Combat Under Environmental Extremes

European and North American armies usually regard normal combat environments as plains or hills in a temperate climate. Mountains, deserts, and the Arctic are usually considered extreme combat environments requiring specialised units, training, and tactics. Although some armies classify jungle as an environmental extreme, the tactical (as opposed to the logistical) aspects of jungle warfare are virtually the same as for forest warfare.

In mountains, steep slopes usually confine even tracked vehicles to the few tortuous roads and make the advance of foot troops away from roads and trails difficult and slow. Constricted routes of advance and the difficulty of deploying forces in mountains have permitted small forces to defend or delay against far larger forces. Leonidas and the Spartans blocking Xerxes and the Persians at Thermopylae in 480 BC is a familiar example from antiquity. In the closing stages of World War II, small bands of Yugoslav partisans delayed the retreat of German and Italian forces across the Dinaric Alps. Larger forces cannot bring their superior numbers to bear against well chosen positions in narrow passes.

Mountains have been a traditional haven for lightly armed forces opposing a heavily armed foe. Unarmoured Swiss infantry could defeat the armoured, heavy cavalry of Austria in their mountainous home. Today, Afghan guerrillas are forcing the heavily mechanised Soviet Army to dismount from its armoured fighting vehicles and retrain in dismounted infantry tactics. The Hindu Kush is not tank country. The problem for medieval knights in Switzerland and the Red Army in Afghanistan was and is mobility. Apart from roads and the flatter valleys, mountains are accessible only to light infantry.

The helicopter may be altering the nature of mountain warfare, but historical horizons are brief. The Syrians used helicopters to seize Mt Herman from the Israelis in 1973, and the Soviets are groping for helicopter tactics against the Afghan guerrillas. An army advancing against an enemy holding a mountain pass had in the past only two options; find another pass or launch a frontal assault. A modern army with helicopters has the option of flying forces over the blocking position. The impact of what the US Marines have called vertical envelopment upon mountain warfare is at present more a subject for speculation than historical analysis.

Deserts have been alleged to permit mechanised forces to operate like ships on a sea, manoeuvring without restraint in any direction. The history of the World War II campaigns in North Africa and the Israeli-Arab battles in the Sinai and Golan reveal a far different picture.

Deserts rarely permit unlimited off-road vehicle movement. In addition to loose sand, abrupt changes in slope restrict trafficability. Erosion in deserts tends to produce steeper and more angular hillsides and stream banks than in more humid regions. These steep slopes and wadi banks create obstacles to cross country movement and channelise avenues of movement. Both the British and German-Italian armies were able to locate favourable defensive positions in North Africa during the 1940s. The extreme fluidity of operations in North Africa was more a function of small armies engaged in battles over large amounts of real estate than of any unusual ease of movement. The angular basalt of the Golan Heights and the mountainous terrain of the Sinai resulted in large scale movements being restricted to existing roads and tracks. In spite of the vast spaces in the Sinai compared to the size of the armies engaged, battles in the 1956, 1967, and 1973 wars tended to be fought over the command of road junctions or passes. In addition to restrictions on movement because of surface forms, the lack of vegetation affects observation and concealment. Vehicular movement raises clouds of dust, and vision and fields of fire are not obstructed by trees, brush, or crops. Arab and Israeli tanks engaged and destroyed one another in the 1973 war at ranges in excess of 2,500 yards. Tank engagements at this range are difficult to envision in the 'normal' terrain of forested western Europe.

Arctic warfare is usually discussed in terms of the severe logistical problems imposed by the cold climate, but the Arctic also has its unique tactical aspects. In the brief summer, mud from the thawing of the ground above the permafrost makes off-road vehicle movement impossible in many places. In winter, the ground and water courses freeze hard and provide for better manoeuverability. The Soviet Union invaded Finland in December 1939, starting the Winter War. Deep snow, however, may restrict off-road movement in the Arctic winter, and in Finland snow kept the motorised Soviet invaders road bound. The defending Finnish light infantry ski units were actually more mobile than the motorised Soviet forces in the snowy forests, and the Finns were able to divide and defeat the Soviet forces in detail. Roads are few and far between in the sparsely settled Arctic, severely restricting mechanised forces when mud or snow makes them road bound. As in mountains, very recent military technology may be increasing mobility in the Arctic in ways that can only be guessed at. The helicopter and the snow mobile may prove superior to light infantry on skis, but they are not part of the normal doctrine of NATO or the Warsaw Pact's tank-oriented armies.

Key Terrain

Key terrain is usually defined as a feature on the battlefield, the control of which gives one military force a significant advantage over an opposing military force. Key terrain obviously varies with the nature of the battleground, the size and nature of the forces engaged, and the level and missions of the commanders involved. Key terrain on a battleground may not even be the same piece of real estate for each of the opposing forces.

In the hilly central highlands of Vietnam, the hilltops are usually grassy and open and the steep hill slopes covered with brush. In the 1960s and 70s, the Viet Cong avoided the open hill tops because of US air supremacy and built their camps and fortifications on the brushy hillsides where they are concealed from aerial observation. Air-mobile US and ARVN units required the open hilltops for helicopter landing zones. Instead of the traditional attack uphill, with defenders on a hill top and attackers attempting to seize it, battles usually opened with a helicopter landing on a hill top and US or ARVN forces assaulting downhill to take a Viet Cong camp on the brushy hillside. The brushy hillsides providing concealment were key terrain to the Viet Cong, while the open hilltop landing zones were key to the air-mobile forces.

For contemporary armies key terrain for division and below is usually high ground within the battle zone. Observation and fields of fire from dominant heights usually permit the occupier a marked tactical advantage over opposing forces. What constitutes high ground, however, varies with the range of the weapons employed. Hannibal ambushed the Romans at Trasimene from slopes no higher than javelins could be hurled or arrows accurately fired. In the American Civil War, artillery and musket fire both had a maximum effective range of about 400 yards. (Cannon could hurl cannon balls much further, but they did not kill many soldiers until they were within range of canister, about 400 yards.) Cemetery Ridge at Gettysburg is actually a rather insignificant rise to a modern soldier's eye, but in 1863 it was an ideal cannon and rifle platform elevated slightly above the Great Valley of the Appalachians. The Battle of Chicamagua's key terrain was desperately fought over Snodgrass Hill, a minor spur of a much larger ridge. Snodgrass Hill is barely noticeable (and unnamed) on modern 1:62,500 maps of that section of the Appalachians, but in 1863 it too was an ideal cannon platform slightly above the valley to the east. The taller ridges were simply too high to fire from effectively with 19th century weapons.

With the vastly increased range of artillery by World War II an observer on a hilltop could call in indirect fire upon most of the targets that he could see. With long range artillery, high ground was valuable not as a weapon platform but as an observation platform. Today, with tank guns and anti-tank missiles that can hit and destroy targets over a mile away, commanding heights are once again valuable as both an observation and a direct fire platform. In the words of the US Army's how to fight manual (FM 100-5), modern weapons are so accurate, long-ranged and destructive, that anything that can be seen on a battle-field can be engaged, and anything that can be engaged can be destroyed.

For echelons above divisions, lines of communication rather than high ground usually define key terrain. Commanders of corps and above are primarily concerned tactically with moving large units about their operational area. Road and rail nets are vital to this activity and key terrain is usually a junction of key routes. Quite often these transport nodes are cities.

Key terrain in harsh environments may be 'refuge' locations. Oases in a desert have been key terrain to both ancient and modern armies. In the cruel winter of 1941-2, Soviet and German small units in northern Russia fought desperately for control of villages and hamlets. A night on the open plains in a winter blizzard could mean freezing to death for the entire unit, and these battles over clusters of huts were hard fought.

Future battlefields, with a dense array of long-range, accurate, and highly lethal weapons, may force commanders to consider areas providing cover and concealment as key terrain. US Army doctrine calls for light infantry to occupy 'tank-proof' terrain when defending against mechanised forces: villages, dense forest, broken topography — any feature which severely limits an armoured fighting vehicle's line of sight, range of engagement, and mobility. NATO helicopter doctrine calls for war-time helicopter flights to be 'nape of the earth'. Helicopters would fly just above the ground and below ridge lines, high bridges, and the canopies of adjacent forests. These measures are considered essential to helicopter survival on a battlefield with modern radar and anti-aircraft guns and missiles. Mechanised forces may also find that a high priority to cover and concealment is mandated if they are to survive on future battlefields.

The Changing Military Geography of Lorraine

The Lorraine Basin of northeastern France forms a corridor between the Paris Basin and the Rhine Valley. Flanked by the Vosges mountains on the south and the rugged Hunsrück uplands to the north it is a well-marched route between northern France and central Germany. Caesar's legions found ancient Celtic forts in Lorraine when they marched east to subdue the Germanic tribes in the Rhine Valley, and Lorraine has been fought over by many armies since then. For present purposes attention will be concentrated on the changes in the tactical military geography of the region during the last three major campaigns fought within it: the opening phase of the Franco-Prussian War in 1870, the Battle of the Frontiers in 1914, and Patton's Lorraine Campaign in 1944. Finally some conjectures will be put forward on tactics in this terrain in a war fought in the 1980s.

During three-quarters of a century, from 1870 to 1944, the geographical face of Lorraine, its hills, climate, vegetation cover and villages, changed very little. This was, however, a period of very rapid change in military technology. In addition, the strategic importance of Lorraine varied greatly from war to war. The same countryside differed significantly in military geography as the opposing armies and their strategic objectives changed markedly from 1870 to 1944.

The dissected topography of Lorraine consists of rolling plains and tablelands, with the latter ranging from isolated buttes to extensive north-south ridges athwart the east-west routes of invading armies. Local relief between upland and lowland ranges from about 300 to 600 feet. The tablelands have flat to rolling upland surfaces and are often bounded by fairly steep slopes, especially on their east-facing flanks. These slopes are often cut by gullies and ravines. Between a quarter and a third of the region is wooded. The forested areas are about equally divided between upland and lowland; tend to occur as woodlots of various sizes, and do not usually form long unbroken belts of forest. Forestry has long been practiced here and the woodlots throughout this period were generally free of undergrowth and had sharp divisions between field and forest. All three campaigns did not begin until August or September, by which time the grains were harvested and forage crops mowed. Except for the minor obstacle effect of vineyards near the Rhine and Moselle valleys, the only vegetation contrast of military significance was between wooded and open areas. The villages of Lorraine usually consist of stone buildings, and villages and forests were alternative sources of cover and concealment during the campaigns.

Figure 5.1: Lorraine

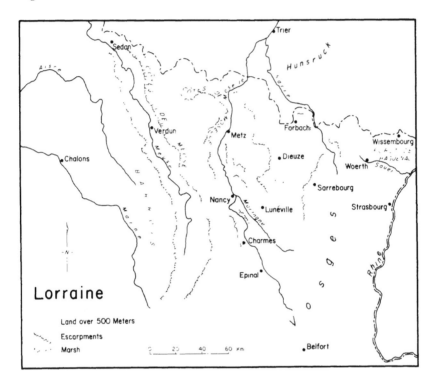

The Opening Phase of the Franco-Prussian War, 1870

Lorraine was the decisive battleground in the Franco-Prussian War. In a series of meeting engagements the Prussians and their German allies dislocated the French armies, setting up the investment of Metz, the French fiasco at Sedan, and the seige of Paris. At the outbreak of hostilities both sides concentrated their armies between the Rhine and the Moselle, along the northeastern frontier of France. The French had better rifles than the Prussians while the latter held the edge in artillery, but the key difference was the superior Prussian organisation. The French were grouped into divisions and corps at the outbreak of war, but not into armies, and they had little practical experience in large-scale manoeuvres. The Prussians and their allies had standing organisations up to field army in size and had conducted realistic, large-scale, peacetime exercises. In most of the engagements fought, the better

Prussian organisation and control permitted them to concentrate superior forces for the battle.

The French had initially planned a defensive disposition, concentrating the bulk of their regular forces at Metz and Strasbourg while their reserves mobilised at Chalons. Napoleon III changed the French plans to an invasion of South Germany as soon as possible, in order to knock South Germany out of the war and to induce Austria to join the French against Prussia. However, the initial defensive-style mobilisation plans were still in effect at the outbreak of war, so the French concentrated at Metz under Marshal Bazaine and at Strasbourg under Marshal MacMahon. The planned French invasion of South Germany was forestalled by the German invasion of Lorraine before the French had fully mobilised.

The Prussians and their allies mobilised in three field armies between Trier and the Rhine, along the northeastern frontier of France. Von Moltke, the chief of the Prussian general staff, wished to crush the French army by a rapid advance southward early in the war. The Germans completed their mobilisation first and by 2 August were marching towards France. The powerful German Second Army in the centre was to pin the French while the smaller First Army, marching along the Moselle, and Third Army, moving on Strasbourg, were expected to fall on the French flanks.

On 4 August the German Third Army pushed a French division out of the frontier town of Wissembourg and moved west along the Sauerbeck after detaching a corps to watch the Forest of Haguenau on its south flank. MacMahon's forces were dispersed for foraging (the French supply system was breaking down) and he called on his scattered corps to gather at Froeschwiller Heights, just west of the town of Woerth, to block the avenue of the German Third Army. On 6 August the Germans closed on Woerth, and although the German command did not intend to launch a frontal assault against the strong French position the leading Prussian and Bavarian units initiated an attack and the rest of the Third Army marched to the guns and entered the fight. French rifle fire took a heavy toll of German troops on the steep, open slopes above Woerth, but Prussian infantry infiltrated a small forest on the French right or south flank and threatened to turn MacMahon's position. French cavalry counterattacked the Prussians in the forest, but their charge was hindered by trees and broken ground and they were cut to pieces by Prussian rifle fire. MacMahon retreated during the late afternoon.

On the same day the First and Second Army failed in a bloody assault up steep slopes against a French corps atop Spicheren Heights, near Forbach. The French, fearful of their flanks, fell back that night from Forbach, moving toward Metz. After Woerth, MacMahon's forces retreated through the Vosges, away from Bazaine's forces at Metz, and toward Chalons. Although the battles of 6 August were tactically indecisive, the two French concentrations were retreating away from each other. Denied a chance to crush the French army near the frontiers, von Moltke seized advantage of the awkward French dispositions and wheeled his armies toward the west.

The Third Army pursued MacMahon, investing Strasbourg with one division and marching unopposed through Lunéville and Nancy. The First and Second Army wheeled on Metz, and while the First Army fought an indecisive battle, just east of Metz, on 14 August, the Second Army crossed the Moselle south of Metz on the 15th. The French attempted to retreat from Metz on the 16th, only to find the road west blocked by a corps of the Second Army. The Prussians occupied a series of unconnected woods atop a low ridge that dominated the undulating plain. In a sharp see-saw battle the Prussians kept Bazaine from breaking through toward Verdun. Both German and French infantry charged through the woods during this fight, but the cavalry, although it would concentrate in woodlots, avoided charging through them.

On the 18th the German Second Army, with part of the First Army, attacked Bazaine's position west of Metz. The French based their position on the village of Gravelotte on the south, where the Germans had to traverse a steep, wooded ravine, and the village of St. Privot on their north flank. On the south the Germans fought their way across the wooded ravine but were repulsed by heavy French rifle fire when they debouched from the woods onto the open upland. On the north the Germans used woodlots for screening their movements and deployed their infantry in skirmish order through the woods. When they could not wrest one small woodlot from the French they forced the French to evacuate it by pouring in artillery fire. The battle around St. Privot went against the French and they occupied the woods of Jaumont to cover their withdrawal. During the 19th Bazaine fell back inside the fortress of Metz where his army was effectively bottled up for the rest of the war. After investing Metz, part of the Second Army joined the Third Army's pursuit of MacMahon's half of the French Army, and the active campaign moved west of Lorraine.

The Franco-Prussian War was primarily an infantry war. Artillery did not yet dominate the battlefield and rifle fire caused over 90 per cent of the casualties. Improved infantry firepower rendered cavalry charges ineffective while the broken topography of the region and its scattered woods made cavalry charges even less feasible. Tactics and organisation on both sides involved modified Napoleonic methods and had not kept pace with improvements in infantry firepower. As a result, the mass infantry assaults characteristic of the battles were generally bloody and indecisive. Armies were still small and compact enough that there were not continuous fronts, and flanks could be turned and obstacles or strong points avoided. The Prussian victories were gained by threatening flanks or seizing key points (such as the Metz-Verdun road) before the French could get enough men on the spot. Although both sides used woods for screening movement or concealing troops, the value of cover and concealment in the face of improvements in firepower does not seem to have been fully realised by either side.

The Battle of the Frontiers, 1914

Although Lorraine was an important battleground in the opening phases of World War I, it was not the only region of conflict and was of course overshadowed by the German advance through Belgium and the 'Miracle of the Marne'. In addition to Lorraine's diminished strategic importance the art of war had greatly altered since 1870. Artillery was the primary killer and, with the machine gun, dominated the battlefield. The size of the armies was so great that they stretched across northern France in a continuous front, and manoeuvre, as it had been practiced in previous wars, was impossible. If a unit had swampy ground or an enemy strong point to its front, it had no choice but to attack through it since the space to its right or left was already occupied by other friendly units. Exposed enemy flanks to turn were rare, since his units usually formed a continuous front from the English Channel to Switzerland.

Once again the French changed their plans from defence to offence just prior to war only to have events force them back to their original defensive posture. After the disasters of 1870 the French General Staff planned, in the event of another war with Germany, to assume initially the defensive from frontier fortresses, then launch a counter-offensive. In the decade prior to 1914, however, French military doctrine favoured the offensive, and a new plan evolved which called for an immediate offensive through Lorraine toward the Saar. If the Germans advanced through Belgium and Luxembourg the French would launch an offensive through the Ardennes simultaneous to their Lorraine advance.

The initial German plan, promulgated by Schlieffen in 1905, called for a massive concentration on the right wing, which was to sweep in a giant wheel through Belgium and northern France. The German left was purposefully weak to entice the French to attack toward the Rhine, away from the decisive sweep through Belgium. After Schlieffen's death the German plan was modified to strengthen the left, or Lorraine, wing.

In the opening phases of the war the French possessed a highly mobile, rapid field gun unmatched by the Germans, but the latter were better equipped with heavy howitzers and had gone further in the development of the machine gun.

On 14 August, simultaneous with a secondary offensive into Alsace by way of the Belfort Gap and the Vosges passes, the French Second Army and the left wing of the First Army attacked toward the Sarre River. German forces fell back from the frontier to strong positions based on the outliers of the Côtes de Moselle, to the east of the river and a marshy area east of these, thus covering the Metz-Strasbourg railroad. The left wing of the French First Army pushed between the Vosges and the marshy area south of Dieuze, reaching Sarrebourg and the trench of the Sarre River, but was repulsed when it tried to cross this trench and forced to retreat. The French Second Army masked the ridge east of the Moselle with a reserve division while one corps attacked along an open valley toward the town of Dieuze. On the Second Army right a third corps struggled through the marsh hoping to debouch from the marsh's northern exits and turn the German position by attacking along the east bank of the Sarre River. The corps in the marsh was split into small detachments by the terrain and could not force the northern exits. The other two corps, in the open valleys, were mauled by German artillery fire, and on 20 August they were counterattacked by Bavarian troops who massed in the forests on the ridge crests. The French, caught in the valleys, could not hold. The French Second Army fell back on the ridges east of the Moselle. The left wing of the French First Army delayed at river crossings and finally stood on the line of the Mortagne River, extending east to the Vosges. The failure of their offensive caused French forces in Lorraine to fall back on their old defensive plans. The defence of France's eastern frontier was based on the fortifications of the Meuse-Moselle line in Lorraine and the Epinal-Belfort line in the Vosges. The Charmes trough between these lines was left unfortified to tempt the Germans to attack along this channelised route, rendering them vulnerable to counterattack from either the Vosges or the heights east of the Moselle.

The German Sixth Army's repulse of the French First and Second

Armies lead the German command to change its role from a defensive to an offensive one. Instead of a single right wing envelopment through Belgium they now attempted a double envelopment by sending the Sixth Army on their left wing on an attack through Lorraine. On 23 August the German Sixth Army advanced south along the Charmes trough. The Germans had fallen for the trap envisioned by the old French defensive plans. German forces in the Charmes trough were raked in the front and right flank by massed batteries on the overlooking heights while the French Second Army vigorously counterattacked their rear and western flank. The German Sixth Army was forced back. Realising that they had to seize the heights east of the Moselle before they could move along the Charmes trough, the Germans next attempted to storm the escarpments north and south of Nancy. These two tablelands have steep east-facing scarps that are dissected into a series of natural bastions and curtains. In most places a fringe of forest atop the scarps screened French batteries and troop movements and concealed French observation posts. The Germans massed large forces in the extensive lowland forests around Lunéville and Nancy and hurled them against the heights to the west, but French artillery broke up the assaults when the German infantry left the forests. French gunners found that the evenly spaced trees along lowland roads made excellent ranging points. After a final unsuccessful attempt on the heights on 8 September, the Germans fell back close to the frontier and Lorraine was relatively quiet for the remainder of the war.

Frontal assaults against enemy positions atop the steep sided tablelands of Lorraine were as ineffective in 1914 as they had been in 1870, only in 1914 the machine gun and improved artillery made them even more expensive. Also, in 1914 there were no flanks to turn while a frontal assault fixed the enemy. The size of the armies, technology of warfare, and the terrain all heavily favoured the defensive in the Lorraine of 1914. Both sides, however, made far more conscious use of forests for assembly areas or for screening troops than was the case in 1870.

The Lorraine Campaign, 1944

The 1940 German invasion of France largely bypassed Lorraine and the Battle of France was, of course, decided by the German push through the Low Countries and armoured thrust through the Ardennes. Intensive campaigning in Lorraine occurred not at the start of the war, as in 1870 and 1914, but near its close, as Patton's Third Army was driving from northern France toward the Rhine. Lorraine was a secondary front, with the main Allied effort occurring north of the Ardennes, and the German winter counterstroke coming through the Ardennes. Montgomery's forces, advancing along the Mauberge-Liege axis on the Belgium plains, directly threatened the critical Ruhr. Patton threatened the Saar, a far less important industrial area than the Ruhr, and Mannheim and Frankfurt, which were of little military importance in 1944.

The advance of the American Third Army from the Moselle to the Sarre in the autumn and early winter of 1944 took three months, while active campaigning in Lorraine in 1870 and 1914 lasted two and three and a half weeks respectively. The fighting differed from both the relatively small number of set-piece battles characteristic of 1870 or the large scale defensive and offensive operations that comprised the Battle of the Frontiers in 1914. Despite the smaller size of the forces in 1944 than in 1914 (Patton's divisions struggled for ground that French and German corps fought over 30 years before) 'battles' in the Lorraine of 1944 tended to be diffuse engagements over broad fronts with many simultaneous, disconnected company to regimental size actions. The continued improvements in weapon range and killing power permitted fronts to be formed with far fewer men per unit area in World War II, than in World War I. In spite of the smaller size of the forces in 1944 the tactical limitations on manoeuvre were similar to those of 1914; there were no flanks to turn and a unit pretty much had to fight through whatever terrain and enemy positions were ahead of it. At this stage of the war Patton's army was much more mechanised than the German defending forces, but the thin road net and poor cross-country mobility within Lorraine gave the Americans few opportunities to exploit this superiority.

Much of the fighting occurred in woodlands, especially where a road ran through the woods or a road junction was located in a forest. The Germans usually mined and blocked roads along American routes of advance wherever these roads passed through forests. American troops became so used to finding the enemy in forests that they routinely shelled wooded areas along their attack routes. Both sides made conscious and extensive use of forests to conceal defensive positions and to assemble forces for attack or counterattack. Toward the end of the campaign the Germans tended to place their blocking positions in the stone villages where shelter from the winter weather was available. American forces found these fortified villages much easier to envelop and cut off than the German forest positions. The hilly and wooded landscape not only broke up operations into a series of disconnected small unit engagements, but also made the employment of armour difficult. The fast-moving mechanised operations that were a Third Army speciality were difficult to conduct. Both American and German field commanders badgered their superior for more infantry. An interesting exception was Delme Ridge, the northwest to southwest trending outlier of the Côtes de Moselle east of Nancy, where an American armoured force stormed its open east slope and overran a German division. What had been a strong defensive position in World War I was taken in a day in World War II, because it was one of the few areas of suitable tank country within Lorraine.

The strategic importance of Lorraine declined during the period studied. It was the decisive battleground of the Franco-Prussian War, an important if secondary front in 1914, and was bypassed in the opening phases of World War II. In 1944 it formed a secondary front to divert German resources from the main Allied thrust north of the Ardennes. The 1870 campaign was the last in the region in which the forces were small and compact enough for classical manoeuvre. It was also the last series of battles in this region where the decisive weapon was the rifle carried by the individual soldier. In 1914 and 1944 fronts were formed and the primary killers were crew-served weapons. The changes in military technology between 1870 and 1914 swung the advantage to the defence. In World War II the tank and mechanisation turned the advantage back to the offence, but the broken topography of Lorraine, with its poor road net and poorer cross-country trafficability, largely neutralised the offensive punch of Patton's armour. Lorraine was no more suitable to armour in 1944 than it had been to cavalry charges in 1870, and all three campaigns were dominated by infantry actions. Another tendency evident in all three wars was for battles here to disintegrate into a confused series of small unit actions as attacking units were divided and dissipated by the broken terrain.

A trend evident during this period was the increasing importance of cover and concealment as weapons improved. The armies of World War II took far more advantage of the woodlands and broken slopes to escape enemy observation and fire than did the armies of 1870. However, very small woodlots in all three wars were traps for soldiers seeking shelter, since they were easily saturated with artillery fire. Also, attacks were frequently broken up when the defender concentrated his fire on the exits from forest assembly areas.

NATO versus The Warsaw Pact — a Scenario for the 1980s

Most published scenarios for an invasion by the Warsaw Pact envision either an attempt to overrun the Federal Republic of Germany in a four to six week blitzkreig, or a grab for both the Federal Republic of Germany and the Low Countries. In these plots the key avenue of advance for the Red Army would be across the north European plain between Hamburg and Hannover. This would provide the most direct route for the Warsaw Pact to both the Ruhr and the Channel ports of the Netherlands and Belgium.

If the Warsaw Pact should attempt to overrun all of Western Europe, a different invasion scene could unfold: a Red Army advance through the Fulda Gap, across the Rhine between Frankfurt and Mannheim, and through the Saar and Lorraine to the Paris Basin. This advance would threaten Paris and industrialised northern France, split western Europe in two, and threaten the Channel ports from the south. This plan could be a tempting one for the Warsaw Pact to initiate if NATO concentrated the bulk of its forces on the North European Plain, gambling on a Soviet main thrust between Hamburg and Hannover and only a secondary Red Army attack in the Fulda Gap.

If a massive Soviet main effort rolled over the US Fifth Corps in the Fulda Gap and forced the Rhine, what sort of engagement could be expected in Lorraine? The Soviet Forces exploiting westward from the Rhine would probably be tank heavy forces, most likely a second-echelon tank arm advancing as fast as possible to exploit the break-through. Most of NATO's mechanised forces, if NATO followed its announced 'forward deployment' policy, would be already engaged both north and south of the Soviet break-through. Forces immediately available to contest a Red Army through Lorraine would most likely be French territorial forces (infantry) and air-mobile forces, probably US and Canadian, from the NATO strategic reserves. How would a French lieutenant general, placed in charge of a hastily formed corps composed of the above elements, fare against a Soviet tank army in Lorraine? He would probably not be too badly off.

The region was not good cavalry or tank country in previous wars. Its broken and forested terrain would force a fast-moving armoured force to stick to the roads. Infantry units could easily block or ambush roadbound columns where roads passed through woodland, forcing the Soviets to halt and deploy manoeuvre and artillery units. A lack of infantry used to dismounted operations would be a further hindrance to Soviet armoured forces fighting through a forest road block. A Soviet tank army in an exploitation west of the Rhine would probably have moved beyond the range of its more sophisticated anti-aircraft weapons and be vulnerable to NATO airstrikes and missile-firing helicopters. The broken and wooded terrain would provide ideal ambush terrain for helicopters firing anti-tank missiles at road-bound columns. Air-mobile units could also ambush supply columns and establish blocking positions along supply routes to the rear of the advancing Soviet columns. A Red Army west of the Rhine in a rapid advance would probably be severely straining its logistical capabilities and air-mobile operations to its rear would be very disruptive. If the Soviets massed mechanised forces against the air-mobile infantry in their rear, these helicopter-borne forces could simply displace to a new blocking position faster than the Soviets could close on them.

The advantages of mobility and initiative would lie with the helicopter-borne light infantry, not with the road-bound army heavy columns. In World War II, dismounted German infantry in Lorraine was able to halt the numerically superior mechanised American forces under Patton. During the 1980s, a helicopter-borne light infantry force with modern anti-tank weapons may well be able to divide and decimate piecemeal a tank-heavy force here.

Readings

Material on the American Civil War is drawn from:
A. Tate *Stonewall Jackson* (University of Michigan Press, Ann Arbor, 1928)
W.B. Woodard and J.S. Edwards *Military History of the Civil War* (G.P. Putnam, New York, 1937)
B.H. Liddell Hart *Sherman* (Praeger, New York, 1958)
V.J. Esposito *The West Point Atlas of American Wars*, vol. 1 (Praeger, New York, 1959)
G. Tucker *Chickamauga* (Bobbs-Merrill, New York, 1961)

Lorraine is dealt with in:
J.F. Maurice *The Franco-German War* (MacMillan and Co., New York, 1899)
B.H. Liddell Hart *The Real War 1914-18* (Little, Brown and Co., Boston, 1930), chs. 2 and 3
D.W. Johnson *Battlefields of the World War* (Oxford University Press, New York, 1921)
H.M. Cole *The Lorraine Campaign* (US Government Printing Office, Washington, DC, 1950)

Forest fighting is discussed in:
J. Miller 'Forest Fighting in the Eastern Front in World War II' *Geographical Review*, vol. 62, no. 2 (1972), pp. 186-202
G.K. Zhukov *Marshall Zhukov's Greatest Battles* (Harper and Row, New York, 1969), ch. 16

US and Soviet tactics are displayed in:
Dept. of the Army *Field Manual 100-5, Operations* (US Government Printing
Office, Washington, DC, 1976)
Dept. of the Army *Russian Combat Methods in World War II* (US Government
Printing Office, Washington, DC, 1950)

6 CAMPAIGN STRATEGY

> Concentrate your forces against the enemy and from a distance of a thousand *li* you can kill his general
>
> Sun Tzu, (chapter 11, verse 57).

Geographic Scope

In the making of war the scale and geographic scope of any decision, order and commitment operates on a continuum. There are no simple breakpoints distinguishing exclusive categories of action. The tactical merges into the strategic, which cannot be untangled completely from the geopolitical. Nevertheless, there is range on the spectrum of responsibility which is concerned with the management of a theatre of war rather than fighting units in a battle and operates to achieve a given political end. Conventionally, this is the function of the general. 'Strategy' conveys the art of disposing, moving and striking with men, machines and firepower to gain a politically determined objective. We have set 'campaign' before it here to emphasise the notion of an organised series of military operations in pursuit of a given goal. In the next chapter we will turn to the geographical designs of policy which guide the conduct of war under the heading of geopolitics and consider the grand strategic perspective which military decisions require.

The geographic resolution of strategy focuses up from the detail of the battlefield to the generalities of the theatre of operations. The concern is with the deployment of divisions rather than individual soldiers, guns or tanks. The movements and locations selected arise from some over-arching design aimed at defeating the enemy. In the words of Liddell Hart the purpose of strategy 'is to diminish the possibility of resistance by movement and surprise'. Clearly, geography enters into the calculations and judgements involved. Movement and surprise bear a dual relationship to one another. Movement creates surprise and surprise generates movement. The possible configurations of this duality are constrained by topography, transport capacity and time. The choice of action when it is made specifies places as well as velocities. It can be pictured as a set of vectors. Geography is not an incidental of the decision but a fundamental variable of an analysis which has the map as its format. Strategy involves a combination of applied geography and psychology. The image employed is of a game played out on a map, shadowing the clash of cunning and force on land, sea and air.

Since 1850 the scale of war has escalated radically along with the scope of economic and political organisation. Industrialisation and the transport revolution unleashed the prospect of total war with massive conscript armies engaged in globally joined conflicts. With this the focus of the critical decision determining the outcome of conflict has risen to encompass a broader and broader scope. Prior to the era of mass armies, once the decision to fight was made, the turning point often lay with the commander on the battlefield. What amounted to a tactical manoeuvre could decide the day and the war. With the coming of railways, telegraphs, long range artillery and massed riflemen, a great deal more organisation and preparation was necessary for fighting. The general staff function emerged, and with it, planned strategies. The increase in scale of operations and the widening of the geographic limits of possible coordinated action raised the level of the turning point decisions to the strategic level. What only the Mongol khans had achieved before in synchronising operations at a continental scale became available to the generals of industrial nations. The talent required to win had to be able to conceive circumstances over a larger geographical extent than that required for the direction of an army on the battlefield. Mechanisation of armies and navies and the arrival of radio communications and aircraft brought the scope of conflict to the global level, and the crucial choices lay at the juncture where the various arms of several nations were coordinated in action. Strategy became subordinate to geopolitical judgement in the pursuit of war, never mind its declaration. The fundamental responsibility for the outcome of war comes closer to the authority which declared it. In the event of war with intercontinental missiles with nuclear warheads, the decision to fight and the responsibility for the outcome rests at the highest level of political authority. The ultimate responsibility for military and political judgement is vested in one person. All else is subordinate to the verdict formed from his conception of the world as it is and might be. However, given the prospect of mutual self-destruction which use of these weapons implies, the use of arms to resolve conflicts persists in indirect competition and limited warfare. And in this setting, strategic calculations obviously play a crucial role, subordinate to the overriding, global agenda. Where the decision to fight a limited war is made, or when the belligerents do not have nuclear weapons, then strategic cunning may be the decisive factor.

The Evolution of Strategic Doctrine

The concept which has enlivened military doctrine over the last half century is mobility. The significance and judicious employment of mobility in war was articulated by the 'apostles of mobility', Fuller and Liddell Hart. Combining historical scholarship with acute observation of the potentialities of new technology and the poverty of military conservatism, these soldiers prescribed the fluid use of tanks and mechanised infantry with air support to overcome the stalemate of trench warfare. Hart preached the virtues of the indirect approach to fighting, drawing upon an historical analysis of military success. Much of what he had to say along these lines had been written down twenty-four centuries before in China by Sun Tzu and it appeals to the intuition of an intelligent boxer, wrestler or football player. The accent on mobility and indirectness of approach was a reaction to the horrors of the muddy grind of World War I, and to seek the genesis of that we have to go back to the beginnings of mass, industrialised warfare.

One of the earliest coherent strategic doctrines operating over a wide geographical domain is seldom remarked by military writers, although it did provide the inspiration for the beginnings of modern geopolitics. Mahan's guide for the US in seeking global power was derived from an analysis of British naval strategy. The fundamental principle of this was the concentration of power. In application this became the doctrine of battle fleet supremacy. Dispersed, hit-and-run operations were a mere supplement to command of crucial waters by a massed fleet which could destroy or bottle up the enemy's forces and blockade their overseas lines of communications.

Napoleon used the increased velocity of the French revolutionary army to achieve a similar concentration of force. In the more uncertain setting of land warfare, he sought to create an unexpected assembly of firepower. His divisions moved over the land, living off it, in 'calculated dispersion' but converged rapidly to join battle, usually in the enemy's rear. As an artilleryman he sought to bring his field guns most effectively to bear by massing them against a particular segment of the enemy's lines. His equivalent of the blockade was the strategic barrage, whereby he manoeuvred onto the enemy's rear and blocked their line of retreat and communications. Napoleon's downfall lay in coming to rely more on mass than mobility in his proclaimed equation of military power with the product of mass and velocity. Along with this he came to substitute a rehearsed battle order for surprise. These errors did not go unpunished in the vastness of the Russian plain, where the defenders merely had to evade the obvious and draw his lines out to destroy his army.

When von Clausewitz put his Kantian generalisations on Napoleon's success down, he emphasised the concentration of forces. Since he expressed his notions in terms of extreme ideals of which reality was a pale reflection, his writing was open to gross misinterpretation. What he intended as the limiting case became, in the hands of some of his disciples, the mode of operation. The aim of strategy was reduced to the destruction of the enemy's forces in battle. The Clausewitzian ideal of absolute warfare was contorted by von Moltke into the doctrine of total war drenched in battles, blood and numbers. The apparent success of this interpretation of von Clausewitz, careful railway timetabling and breech loading rifles in 1866 and 1870 by the Prussians, established the doctrinal orthodoxy of the direct approach.

On the other side of the Atlantic, Grant's victory in a head-on war of attrition against the Confederacy seemed further vindication of directness. Yet Grant had learnt the value of mobility and manoeuvre in the west, especially in his capture of Vicksburg. There are those who credit Sherman's extension of this western style of mobile war with the eventual Union victory. The dependence of massive conscript armies on railways for their transport and supply reduced the fluidity of war, giving a very limited range of operation for rail fed armies in terms of distance from rail lines. These lines also posed a problem of vulnerability, providing distinct targets for disruptive raids or the establishment of a Napoleonic, strategic barrage. On the Confederate side, Forrest and Morgan demonstrated the efficacy of line-cutting raids. To win Vicksburg, Grant captured the railway junction at Jackson as a barrage cutting the Confederate line of communication eastwards. Sherman, appreciating the vulnerability of fixed supply lines and the advantages of mobility, deliberately cut his army's dependence on railborne supplies and struck at the enemy's rail dependency. His goal was not confrontation with their army, but control of crucial lines and junctions in their transport system, starting with Atlanta. From thence he marched to the sea with what amounted to five flying columns, endangering a multiplicity of strategic points and, thereby, throwing the Confederates into confusion over his expected line of advance. His ability to change direction at will eased the task of cutting Richmond off from its source of supply in Georgia. On reaching the sea, he turned left and proceeded to roll up the South's ports, its vital connections to overseas supply. It was this campaign which seems to have most vividly impressed the value of mobility on Liddell Hart, thus it planted the seed of future theory and its translation into practice by Guderian's panzers.

It is ironic that in preparing for World War I, France should have had a plan based on the doctrine of seeking the single decisive battle derived from the Prussian von Clausewitz, while the Germans' intended strategy was fluid and sweeping, redolent of early Napoleonic victories. France's Plan XVIII called for a united blow against the centre of the German line. The original plan of von Schlieffen was to wheel around the French wall through Belgium with a major part of the force available on the right wing, rapidly engulf the Paris Basin and, coming up on the French in their rear, hammer them into submission against the anvil of the German fortresses in Lorraine. A lack of geographical boldness led von Moltke to modify this, strengthening the left wing facing the main French force, at the cost of weakening the right and, thus, the speed of its advance. The Allies held the western swing in Flanders, leading to stalemate and the war of entrenched attrition. British efforts at outflanking Germany on a continental scale by driving through the Balkans after forcing the Dardenelles, employed British command of the sea imaginatively but failed in the hesitancy of its execution. It was this British command of the sea which, in denying material and food to the economic base, eroded Germany's military resistance.

The success of unorthodox methods in the Middle Eastern side-show with the Arab revolt, so ably publicised by Lawrence, and Allenby's thrusting defeat of the Turks in Palestine, was further evidence of the value of indirectness and mobility. In the closing stages of the war in Europe, the tank and aeroplane's potential for breaking the rigidities of trench warfare were demonstrated. The bloody, muddy horror of unimaginative adherence to Clausewitzian doctrine lay strewn across the scarps and vales of northern France. All of this was taken to heart by Liddell Hart in formulating a new strategic theory.

Mobility and the Indirect Approach

In the early 1920s Fuller and Liddell Hart preached the virtues of deep penetration of enemy formations with tanks. They differed in that Fuller saw infantry's only function as holding strategic points, whereas Liddell Hart would have foot soldiers carried along with the armoured drive. In 1927, the British Army formed an experimental mechanised force to test the words of these prophets. In Germany, von Seeckt had seen the need for increased mobility but only in terms of mounting troops in trucks. Guderian picked up the British authorities' emphasis on tanks and aircraft and moulded them into the panzer division. The strategy and tactics for their employment in the blitzkrieg were provided by Liddell Hart's formulation of the 'expanding torrent' method of attack. He saw the aim of strategy as dislocation. Force should be deployed so as to throw the enemy into turmoil, maximising his uncertainty and disrupting his lines of communication, supply and retreat, damaging his capacity to resist physically and psychologically. The tactics to achieve this involved a surprise breakthrough with planes and tanks to cut a gap in the enemy's front. A torrent of tanks and infantry, carrying supplies to give them a long range of uninhibited action, is poured through the gap to expand behind the front into fast independent units thrusting along lines of least resistence, swerving around opposition, speeding up where a passage of advance is narrowed so as to maintain the force of the thrust, and selecting a sequence of objectives in such a manner as to keep the opposition unsure and unbalanced.

The devastating economy of this was demonstrated by the panzers in Poland in 1939 and France in 1940. The tank tide was decanted through the Ardennes to drive deep and far, separating and rolling up the French and British. For the German attack on Russia, the ratio of opposing force to space was so low on the plains that the large array of possible lines of advance allowed the tide to roll rapidly to the Volga. The Allied defenders took the benefits of mobility to heart also. The tide was turned in North Africa with a brilliant display of the mobile, indirect approach by O'Conner in the Western Desert. Montgomery, trained in the old style of strategy, learned the profitability of Liddell Hart's doctrine and employed it with success. Patton's break out from Normandy and sweep to the Marne was a classic exercise in *l'audace* and the indirect approach, until he was drawn into a frontal attack on Metz.

The spirit of Mongol mobility which Tukhachevski had injected into the Red Army was lost when he was shot along with much of the upper echelon in Stalin's purge of 1937-9. However, the geographical setting, Russia's vast and open plain, made a strategy of mass and manoeuvre, such as he had prescribed, the inevitable choice. Tukhachevski had left a legacy including training in rapid encirclement operations; his 1936 Field Regulations emphasising speedy offence combining infantry, tanks, artillery and aircraft; and organisation of assault groups to breach enemy lines and provide the gap for a torrent of tanks. These combined with greater numbers and space to beat the German armies. As the Wehrmacht retreated, the disadvantages of being spread along an extensive front were reduced, but manpower was reduced more than proportionally. Even so, German mobility proved its worth in a flowing defence which beat off attacks of more than ten to one against.

The establishment of the US Army was distinctly Clausewitzian, bent on seeking a sharp and decisive, head-on confrontation with the enemy and this was a point of contention with British strategy. The Pacific theatre was essentially a naval show, employing the Marine Corps' amphibious landing tactics to hop from island to island, bringing air strike capability closer and closer to Japan. The naval strike across the central Pacific was complemented by MacArthur's brilliantly devious advance on Japan via New Guinea and the Philippines in a strategy of opportunity.

Airforce Strategy

Trenchard, the commander of the Royal Flying Corps in France in World War I, who became the Chief of the Air Staff, was convinced by 1917 of the offensive value of the aircraft in long-range, 'strategic' bombing operations. Through the 1920s and 1930s he demonstrated the versatility of aircraft in a series of operations in the Middle East. He propounded a thesis that air power was far more effective than ground forces in carrying out the imperial function of territorial control. The allocation of resources to the growth of the RAF between the wars was a political response to popular fear of a direct air attack on British cities. The expenditure was governed by a rule of parity with the air force of any country within striking distance. Trenchard assured that the expansion was weighted heavily in favour of bombers for the offensive. He saw air defence as attack. In Germany, Goering's Luftwaffe was built to play a key role in achieving the demoralisation and disorganisation which attended a blitzkrieg.

The literary groundwork for strategic bombing was laid by Douhet and Liddell Hart. In 1921, Douhet, an Italian general, dismissed the potentialities of mobile ground warfare and proposed that war be transferred to the air. Aircraft could be used to terrorise the civilian population causing political disintegration and, thus, military collapse. Writing in 1925 Liddell Hart had proposed that a mobile armoured lunge against the enemy lines be compounded with an air strike at their economic and political system, bombing civil centres. Trenchard adopted this as a text for the development of a new form of war of attrition. This was to take the form of an erosion of industrial capacity rather than the mass wasting of trench warfare.

Trenchard's aggressive emphasis on long-range bombing was carried back to the USA by Mitchell. His advocacy inspired the instructors of the Air Corps Tactical School of Maxwell Field, Alabama, who were destined for commanding positions in World War II. They taught that the object of the airforce was to break the will and power to resist of the opposing nation, aiming at the productive capacity which sustained military strength. The geographic focus of attention was shifted from military formations and defences to the location of population, industry and transport facilities. These apostles went far beyond Mitchell in geopolitical terms, for they foresaw a need not merely for off-shore bases directed at potential enemies, but for arrangements with likely allies to locate bases to extend the range of US air power. The American tactic of daylight precision bombing came to grief in the mess of the Schweinfurt raids in October 1943. The British policy of night time, saturation bombing of residential and industrial areas instituted by Churchill in May 1940, also proved a failure. It did not prevent the production of war material expanding apace and, if anything, it enhanced the morale of German civilians. In March 1944, Tedder's proposed use of bombers to demolish the French and Belgian railways was employed successfully to reduce German mobility in preparation for the Normandy landings. The decision to drop atom bombs on Hiroshima and Nagasaki in August 1945 was based on domestic political needs and the desire to outpace the Soviet Union in global influence. The evidence suggests that the Japanese were willing to surrender on essentially the same terms as the final, qualified 'unconditional surrender' as early as May 1945. Certainly, the Emperor was convinced of the need to end the war by June and was seeking Soviet mediation from that time. In retrospect, the use of these weapons looks more like an experiment in wielding power than a military necessity.

Nuclear Strategy

The course of the war and the methods employed by the Allies left an impression of the superiority of offensive ingenuity seeking fast and total victory. Amphibious assault, carrier task forces, tanks and strategic bombing seemed to have brought success. Even before the final surrender of Germany and Japan the lines for the next confrontation were forming and plans for its prosecution were being drawn up. The invention of nuclear weapons overturned the balance of violence, giving

the overwhelming advantage to the offence. In 1948, Bradley enunciated the doctrine of 'massive retaliation'. Defence was to take the form
of the threat of enormous, offensive destruction. As a counter to any
aggressive Soviet move, the US stationed its forces so that it could
strike the Russian homeland with nuclear bombs carried in B36s and
B50s. A small deployment of conventional forces would provide the
'trip wire' for this response. Stumbling against this the enemy's aggression would automatically bring destruction down on their own heads.

In 1949 the USSR tested their first A bomb and in 1955 their first
H bomb. With the Soviet acquisition of these weapons the protagonists
faced each other in the 1950s with the capacity to deliver nuclear
bombardments across the oceans and massive retaliation was no longer
credible.

From the Soviet viewpoint it appears as if the army in eastern
Europe, which the US regarded as an offensive threat destined for
global conquest, was considered by Stalin as a deterrent to point at
western Europe, counterbalancing the US monopoly of nuclear power.
In 1953, Krushchev rejected as too risky the strategy of a preemptive
strike to destroy the US threat before it could get off the ground. The
strategy for Soviet ground forces was essentially similar to that which
NATO came to adopt, the forward active defence. Greater mobility
and firepower, including tactical nuclear weapons, were employed to
allow for speedy thrusts along widely separated axes. Medium range
missiles were incorporated in the armoury to hit European targets.
Lagging behind in ability to deliver nuclear bombs the USSR began to
build up its long range bomber fleet of Bisons and Bears as well as
developing anti-aircraft defences to protect its cities.

In the USA fear of a 'bomber gap' in the mid 1950s led to the B47
and B52 programmes. In the meanwhile the USSR had decided to neg-
lect long-range bombers in favour of rockets. The launching of Sputnik
in October 1957 signalled their success and by 1960 the SS6 was operational. In 1959 Kruschev perceived the magnitude of destructive power
to be roughly in balance, having been won by the Soviet development
of intercontinental rockets. He established the Strategic Rocket Force
as the spearhead of deterrence, which would allow a reduction of
conventional forces.

In the USA, this was viewed as a dangerously unstable state of
affairs by the military establishment with the 'missile gap' inviting a
Soviet surprise attack. Even though President Eisenhower did not think
the Soviets had enough force to knock out the US bomber fleet, the
pressure was strong enough to get the Atlas, Titan, Minuteman and

Polaris programmes underway. When US satellites revealed in 1961 that the USSR had only sited a few ICBMs near Moscow, and evaporated the misperception of a missile gap, it was too late. US rocketry soon left the USSR far behind. The prospect of being outgunned by Polaris and Minuteman was the strategic basis of Krushchev's effort to outflank geographically the US early warning network by placing missiles on Cuba in 1962. Faced down over this, the Soviets turned to diplomatic means to decrease the tension, seeking detente while countering offensive inferiority by building up their air defences and anti-rocket systems.

From the time the Soviets achieved 'second strike capability', that is, the theoretical capacity to absorb a US attack and strike back, massive retaliation gave way to 'flexible response' as the strategic doctrine of the USA. This embodied the notion of achieving victory in a nuclear exchange by developing the capacity to destroy the USSR, China and their satellites as national societies after withstanding the worst possible attack.

It was, however, becoming clearer that there was no real hope of a defence against nuclear attack. The offence can pick the time, place and magnitude of the first blow. After this it would be a matter of savaging each other until one side's supply of weapons is exhausted. In such an affair there would be no victor. By the 1962 Cuban missile crisis, the USSR and USA had enough firepower to smash each other beyond repair. In realisation of this, nuclear weapons came to be viewed as the ultimate deterrent. A surprise attack with nuclear weapons would be fended off by the certainty of retaliation in kind. A large scale strike with conventional forces in Germany would be held in check by fear of the inevitable stepping up to nuclear war. This restraint through fear had to be reciprocal to work, so that neither side had the slightest temptation to strike out first. The conception arose of global aggression held in check by the horror of 'mutual assured destruction'. Each side was in a position to come back after the first blow and savage the attacker sufficiently to turn any hope of victory to ashes. For this balance to be struck, it is in their mutual interest that each side be perceived to have an invulnerable retaliatory force. This removes the temptation to strike first. If either side is worried about its vulnerability to an unexpected blow, there is the danger of adopting a 'launch on warning' policy, raising the spectre of accidental war.

Lagging behind in offensive power from the outset, the USSR had concentrated on defences — radar, surface-to-air missiles and interceptors. In the late 1950s they developed ABMs (anti-ballistic missiles) some of which were installed around Moscow by the mid 1960s. Realising the impotency of ABMs by 1962, the Soviets began to express interest in arms limitation. Despite evidence of the ineffectiveness of ABMs, the US began to develop MIRVed weapons to overwhelm them. At the same time that Nixon and his Secretary of Defence, Melvin Laird, were saying in 1971 that the US was not developing a first strike force and would not make any moves that even looked like it, they were having MIRVs deployed.

The MIRV upset the deterrent equilibrium because each missile could now kill several on the ground. As long as a missile had only one warhead, given the probability of failure, there was never an advantage in using one missile to attack another. The exchange ratio, the number of kills per missile, would always be less than one. The MIRV reverses the disadvantage of the attacker if accuracy and reliability are high enough to give each warhead a reasonable chance of destroying a missile silo. The exchange ratio becomes favourable to the aggressor. The US first tested such weapons in 1968 and had them ready for action in 1970. This was bound to draw a response. The USSR tested its first MIRV in 1973 and had them sited on land by 1975 and in submarines by 1979. By 1980 the Soviet MIRVed missiles were considered such a threat to US land based ICBMs, that the MX was called in to close the defensive 'window'.

There was a radical departure from the attitude of reciprocal balance in 1973 when Schlesinger replaced Laird and the notion of a winnable nuclear war began to take shape. In September 1974, Schlesinger laid the groundwork for this idea when he ran down US preparedness and overstated Soviet potential before the Senate Foreign Relations Committee. He presented a scenario in which a 'surgical' Soviet first strike would knock out US silos, killing only 800,000 civilians. The president would then capitulate for fear that any counterstroke would be met by massive destruction of urban America. Nuclear war was thinkable, its outcome acceptable and the silo-busting MX had a justification. A more reasonable estimate of the dead from such a first strike is 20 million. This would clearly pose the risk of anguished and awesome US response, especially since the other two legs of its deterrent triad would be virtually unscathed. The decision to deploy the MX in existing silos rather than indulge in any elaborate locational subterfuge runs the danger of looking more like the mounting of a first strike weapon to the Russians.

Guerrilla Warfare

Nuclear weapons introduce a fundamental discontinuity into the calculus of force. They give their owners enormous superiority in potential destructive power over those without them. However, when their use would involve a step up from imposing a loss of thousands of people to the obliteration of their society on an opponent, to use them as a threat becomes unreal. Since it might also cause the other member of the nuclear duopoly to respond in kind, the risk of escalation makes any threat even less likely. Thus, the possession of nuclear weapons alone becomes irrelevant in the shatterbelts between the duopolistic competitors, where their use would be stupid. Denying an enemy territory by reducing it to sterile ashes is hardly victory. Superior firepower alone has proven incapable of winning the guerrilla wars which have proliferated in the shadows between the nuclear titans. We will discuss the tactics and strategy of both sides of guerilla war and its geographical setting in chapter 8. In strategic terms it clearly occupies a niche created by the hiatus between unemployable nuclear superiority and the willingness and capacity to fight on the ground on the part of major powers. In this light it is not surprising that one of the chief sources of intellectual ferment in the Pentagon currently is the message preached by John Boyd that successful warfare is psychological rather than physical. The way of the future is seen as residing in the mobile and deceitful search for an enemy's weakness rather than fighting his strength in big battles. Sun Tzu would seem to have had the right idea.

Readings

British naval stratety was analysed by:
A.T. Mahan *The Influence of Sea Power upon History, 1660-1783* (Little, Brown, Boston, 1890)

The apostles of mobility can be consulted in:
B.H. Liddell Hart *Strategy* (Praeger, New York, 1967)
J.F.C. Fuller *A Military History of the Western World* (Funk and Wagnalls, New York, 1956)
K. Macksey *Guderian: Creator of the Blitzkrieg* (Stein and Day, New York, 1975)

To illuminate airforce and nuclear strategy there are excellent essays by Gibbs, Matloff, Mackintosh, Kissinger and Buchan in:
M. Howard (ed.) *The Theory and Practice of War* (Indiana University Press, Bloomington, 1965)
We have already referred to a particularly well-informed volume on nuclear strategy and logistics in Scoville's *MX: A Prescription for Disaster.*

7 GEOPOLITICS AND GRAND STRATEGY

> When a state is enclosed by three other states its territory is focal. He who first gets control of it will gain the support of All-under-Heaven.
>
> Sun Tzu, (chapter 2, verse 6).

It is difficult to determine the extent to which geopolitical ideas have directly influenced policy and action, or whether they merely reflected the mood of the times. Whether the spark of decision came from the theory or whether the theory merely rationalised the whims of rulers is not always clear. But there are strong grounds for suspicion that the written word and eloquent expression of geographical generalities have encouraged certain tendencies in political choice which translated into violent action.

Mahan and US Strategy

From the Tudor period successive British governments pursued a coherent geographical objective by diplomatic and military means. They endeavoured to keep the mouth of the Rhine, facing them across the North Sea and focusing the Atlantic face of Europe, out of the hands of a land power, be it Spain, France or Germany. The military means to this end was a navy which could take on those of any two potential rivals. This was disposed so as to control the seaways in and out of Europe. The beginnings of US overseas imperialism, the extension of manifest destiny to a global setting, were strongly affected by Captain Alfred T. Mahan's study of the influence of British sea power on history. In 1890 Mahan pointed out that British supremacy rested on the Royal Navy's control of the Eurasian balance of power. America's best chance of achieving security and greatness lay in complementing the British with a battle fleet and forward bases for their ships, commanding the Pacific. The inevitable competitor in this ambition was Japan. Through the 1890s Mahan preached the need for a chain of bases from Cuba through Panama, where the canal was being cut, to Hawaii. His ideas impressed members of the ruling class, such as Henry Cabot Lodge and Theodore Roosevelt, who became advocates of his doctrine. More widely, the prospect of new markets and missionary fields attracted business and clerical support. The trigger to action was the 1895 rising against Spain in Cuba. The expansionists extended Mahan's vision to include the Philippines, with his approval, calling for an expeditionary force to plant the stars and stripes on Filipino soil.

After the excursion into the European theatre in 1917-18, US global strategy reverted to an essentially one-ocean, naval stance, leaving US interests in the Atlantic to the Royal Navy's care. In the early 1920s military planners devised the Orange plan for a naval war with Japan. By the mid 1930s the planners decided that Japan could only be defeated at great cost including the initial surrender of the Philippines. The divergent aims of the services led to a compromise final plan in 1938. This combined the army objective of concentrating defence on the 'strategic triangle' of Alaska — Hawaii — Panama with the navy's predilection for an offensive war, island-hopping across the Pacific with carrier task forces and amphibious operations.

The simplicity of this attitude was upset by the events of 1939-44 which switched attention from the Pacific to the Atlantic, and called forth a swatch of so-called 'rainbow' plans involving war with combinations of enemies and allies. The geographical scope of US defensive strategy extended its perimeter from the continent to the hemisphere, on the assumption that British survival was vital to US security. Despite the pleading of MacArthur, General Marshall the Chief of Staff, came to the conclusion that the Pacific position should be purely defensive, with the main force in the Atlantic. This view was embodied in Plan Dog at the end of 1940, involving large land operations in Europe in close cooperation with the British. In the events that followed, the simple thread of geopolitical doctrine became tangled in the complexity of action.

Mackinder and Geopolitik

The source of another simple but influential view of the world lies in the work of Sir Halford Mackinder. In 1904 he identified the central and northern plains of Eurasia, inaccessible to ships but amenable to the drive of cavalry or railways, as the 'pivot of history'. From this position, regardless of her constitutional and social make-up, Russia, as the successor to the Mongol Empire, exerts the primary political pressure on the globe. This pressure is transmitted to four marginal divisions of Eurasia where naval power can be exerted. These peripheral regions were east Asia, south Asia, the Middle East and Europe. After World

War I he revised his thesis and retitled the Russian pivot the 'heartland'. This was fringed by an inner crescent, comprising the four marginal regions previously identified, and an outer crescent consisting of the British Isles, Africa south of the Sahara and Japan. The fundamental opposition arising from this array of landmasses confronts the land power of the heartland with the sea power of the crescents. The object of the revision was to warn British statesmen of the danger to Britian's maritime empire of the combination of Germany and Russia. The epicentre of political strife was shifted west to eastern Europe. Whoever combined the heartland with Germany would rule the world island, Eurasia. Whoever ruled Eurasia, ruled the world.

The seedbed for the reception and employment of such ideas had been prepared in Germany by a tradition of political geography stemming from Ratzel in the 1880s. The term 'geopolitics' to describe the use of geographical concepts to serve the state was coined by a Swede, Rudolf Kjellen. The chief apostle of geopolitics was Karl Haushofer, a Bavarian military aristocrat who had been an observer for the general staff in Japan and Southeast Asia in the 1900s. During World War I Rudolf Hess was his ADC. After the war Haushofer campaigned for the recognition of geopolitics as a powerful instrument of state and in 1924 started what was to be a widely circulated magazine, *Zeitschift für Geopolitik*. Haushofer's ancestral home outside Munich was a refuge for the embryonic Nazi party and he was a frequent visitor to the imprisoned Hitler during the period in which he dictated *Mein Kampf* to Hess. There is certainly a leavening of geopolitics among the racist hate which dominated Hitler's utterances. Haushofer and the majority of the geopolitical writers remained aloof from racism. He stood by his Jewish wife and this is cited as the reason for his lack of elevation in the Nazi ranks, in which he aspired to be minister of education.

Mackinder's writings provided Haushofer with a specific world view. What was a warning of danger to Britannia's rule of the waves became an invitation to seek an accommodation with Russia to defeat the Anglo-Saxon monopoly of power, including the USA in this domain. In 1913 Haushofer wrote that a community of interest between Japan, Russia and 'the central European imperial power' would be absolutely unassailable. The Nazi agenda for global domination was a flexible plan directed at two underlying geopolitical objectives. Firstly it was necessary to obtain control of the heartland. Then it was necessary to destroy the naval power of the UK and USA. This is foreshadowed in *Mein Kampf*, echoing Haushofer's use of Mackinder's conception.

Haushofer worked consistently for an accommodation with Russia. The Hitler-Stalin Pact of August 1939 he saw as a vindication of his basic contention, providing an axis from the Rhine to the Amur and beyond to Japan. To Hitler, it was a temporary means of immobilising Russia while he was taking Poland. Although Haushofer's editorials after Hitler broke the treaty in 1941 were patriotically proper, it is clear that he regarded it as a geopolitical mistake, pointing out the difficulties that Napoleon and von Falkenhayn had in combating the vast spaces of the heartland. The prospect of mutual combination which Mackinder had feared and Haushofer campaigned for was dashed by Barbarossa. No geopolitician had foreseen the configuration in which British and American seapower was allied with Russian land power. Hitler evidently had wished to avoid getting embroiled in a battle with British seapower. He persisted in regarding the continued existence of the British Empire as an indispensable part of the world order. The geopoliticians foresaw its demise and predicted the transfer of leadership to the USA. Obviously, Hitler's racialism overwhelmed the influence of the geopoliticians. If this had been stronger, Germany might not have been so vulnerable.

Shatterbelts and Dominoes

In the aftermath of World War II, the new political alignments and lowered technological limitations led Saul Cohen to describe the geography of world politics in terms of cores and shatterbelts. In 1915 Fairgreve used the term 'crush zone' to describe the tier of small states lying between the heartland and the sea powers. Writing in 1942, Whittlesey designated the buffer of small states lying between Germany and Russia, which had been identified cartographically by the German geopoliticians in the 1930s, a shatterbelt. Any array of cores and shatterbelts identified obviously represents an impermanent state in a dynamic process. The forces of conflict resolve themselves in sporadic rise and fall of cores and the consolidation and fragmentation of shatterbelts.

One particular potent image conjured up to dramatise this competition is the so-called 'domino theory'. The attitude that this characterisation of political geography conveys seems to have coloured many diplomatic and military decisions by the USA over the last quarter of a century, and the image has been used in the utterances which supported such actions. The domino attitude is alive and well. During the 1980 presidential campaign, Ronald Reagan said 'Let us not delude ourselves.

The Soviet Union underlies all the unrest that is going on. If they weren't engaged in this game of dominoes, there wouldn't be any hot spots in the world'. Secretary of State Haig constantly referred to the sinister influence of the Soviets as the prime mover inspiring Cuban assistance to guerrillas in El Salvador, Colonel Quaddafi's efforts to merge Chad with Libya and SWAPO's drives into Namibia from Angola. In the spring of 1982, with panic over Central America, the domino theme was resurrected explicitly and a falling domino motif decorated *Time* cover stories. Henry Kissinger's apologia, *Years of Upheaval*, was published at this time and he endorsed the theory: 'The impact of a North Vietnamese victory on the prospects of freedom and national independence in Southeast Asia was certain to be grave; the much maligned domino theory turned out to be correct'.

The fear which underlies this view of the world was spawned in the 1940s when a sequence of Mediterranean countries was perceived to be tumbling into the Soviet sphere in the aftermath of World War II. The Truman Doctrine was promulgated as a counter to this. This alarm was articulated and the extent of the globe endangered was expanded by William Bullitt, a former US ambassador to Paris and to Moscow, in an article published in *Life* in 1947. This voiced the fear of monolithic communism emanating from its Russian power source and engulfing the world via China and Southeast Asia. This was a new political version of the threat of the Asiatic horde flooding off the steppes to engulf the civilised world, which finds expression in the writings of Major General Fuller, for example. This is compounded with the myth of Russia's methodical policy of conquest, dating back to Peter the Great. A premier objective of this aggression was, according to the myth, a warm water outlet. However, the first Russian war with the Ottoman Empire against the Khan of Crimea in 1688, under the Regent Sophia, was driven neither by the desire for access to the sea nor a holy crusade to free Constantinople, but was rather the reluctant satisfaction of a treaty obligation to Poland to ensure retention of Kiev. Whatever the motive, the geopolitical effects of this campaign still resound far afield. This is the genesis of the boundary dispute with China and the continuing confrontation across the Amur. The forces employed against the Crimean Tartars were denied to the Russian advance towards the Pacific. Negotiations with China under duress produced the Treaty of Nerchinsk giving the whole Amur Basin to China. In the 1850s Russia took it back claiming that the treaty was invalid and the dispute continues.

The dynastic ambitions of the Romanovs, or the crusading spirit of orthodoxy, or the racial imperative of the Russ, or the heritage of fear from the fall of Byzantium which was Russia's birthright, could be replaced after 1917 by the missionary zeal of revolutionary Marxism. This provided a new mythical driving force to power the expected outpouring of the Asiatic horde. Frequently, the myth is presented in compound form. In its Southeast Asian extension, the spread of communism was tied in with Han ethnic imperialism. In 1954 H.J. Wiens, a geographer, produced a scholarly version of Bullitt's justification for American intervention in Vietnam. The 3,000 years of Han expansion southwards from the banks of Yangtze provided an inexorable process which the Soviet strategists could harness for a political and military assault on the colonial powers in order to create a new communist empire. Scholar to the end, Wiens maintained an academic ambivalence as to whether the pressure on South Vietnam, Cambodia, Laos, Thailand, Burma and Malaya represented the historical momentum of the Han or a new force fuelled by Soviet Russia.

It was Admiral Arthur Radford who first used the analogy of a row of dominoes in 1953 when he was urging a carrier based, nuclear bombing strike to relieve Dien Bien Phu in a meeting of the Joint Chiefs of Staff. Eisenhower took up the catchy phrase immediately, suggesting that the loss of Indochina would cause the call of Southeast Asia like a set of dominoes. Next month, when direct military intervention no longer looked attractive, Eisenhower and Dulles both denied the veracity of this image, claiming that the rest of Asia could be held even if Indochina fell. So, from the outset of its currency there was doubt about the validity of the concept. The seed had, however, been planted in the rhetoric of US officialdom, along with the impression of a friendly South Vietnam government as a strategic necessity for the USA. It was around this time that the domino model made the transition from simile to theory.

Walt Rostow and Maxwell Taylor were the chief purveyors of domino theory to the Kennedy administration, converting McNamara to this view. Within months of his inauguration, Kennedy was elaborating on domino theory, suggesting the communist control of Laos would jeopardise the West's strategic position in Southeast Asia.

The opponents of the theory attempted its demolition by pointing out the greater reality and urgency to the local people of long-standing rivalries between Burmese and Thai, Khmers and Thai, Khmers and Annamese, Malays and Javanese, Filipinos and Indonesians, rather than the ephemeral clash of communism and anti-communism. The source of unrest in the region is the deeper sense of nationality which is inflamed by any foreign presence. Therefore, intervention based on domino theory generated the very forces it hoped to contain. The significance of China in the region is a matter of geographical fact rather than ideological geostrategy. The assumption of China's desire to expand her bounds beyond the imperial limits can be seriously questioned.

What brought the domino theory into question in terms of the train of events was the reversal of communist fortunes in Indonesia after the death of Sukarno in 1965, when party members were savagely massacred. What was regarded as a falling domino was stood up without direct US intervention. In May 1967 McNamara was convinced of the error of the domino doctrine and attempted to defuse it. In broaching the prospect of a more politic, less militarily quantitative approach with President Johnson, McNamara cited the Indonesian case and pointed to the first stirrings of Chinese realignment. The idea did not, however, succumb to these attacks and was inherited by Nixon. In an interview published in *The Times* in June 1970, Nixon said

> Now I know there are those who say, 'Well, the domino theory is obsolete.' They haven't talked to the dominoes. They should talk to the Thais, Malaysians, to Singapore, to Indonesia, to the Philippines, to the Japanese, and the rest . . . and if the United States leaves Vietnam . . . it will be ominously encouraging to the leaders of communist China and the Soviet Union who are supporting the North Vietnamese. It will encourage them in their expansionist policies in other areas.

Not only did the currency of the theory survive into the Reagan administration but the field of its incidence was extended into Africa, Europe, and Central America. The domino effect was deemed to leap oceans and continents to start up again in Libya, Tanzania, Nicaragua and Italy. Proponents of the theory predicted the fall of Thailand. In 1976, a group including Kissinger, Ford and John Connally publically urged Italian leaders to keep communists out of government in Italy to avoid endangering the security of the entire Mediterranean.

The Southeast Asian case, however, is still the ultimate test of the model's veracity. An examination of the train of events of the last decade suggests more complex causes and motives are involved than those postulated by this seemingly powerful theory. It was US action which brought Laos and Cambodia directly into the fray. The US invasion of Cambodia in 1970 shook Prince Sihanouk loose in favour of Lon Nol, whose weak grip was replaced by the excesses of Pol Pot and the Khmer Rouge regime in 1975. The Khmers Rouges, with their draconian policy of dispersal of city dwellers, were clients of the Chinese and at odds with Hanoi and the Viet Cong, by now enjoying Russian support. In 1977 the victors of Vietnam swept into Cambodia and captured Phnom Penh. The refugees in Northern Cambodia and over the border in Thailand, created by Pol Pot's actions, were now infiltrated by remnants of the Khmers Rouges. In retaliation for the Vietnamese action China, by this stage having achieved a *rapprochement* with the US to balance its antagonism to Russia, mounted a punitive attack on Hanoi. In mid 1981 Vietnam's occupation is faced with three Cambodian resistance groups, the Khmers Rouges, a rightist movement led by Son Sann and one loyal to Sihanouk, who is negotiating with all parties involved, including China. As fear of a Vietnamese drive into Thailand and Malaysia evaporates with their poor military and economic showing, the geopolitical value of a stable Vietnam as a bulwark against the awesome presence of China is coming to be valued by the ASEAN nations. The resiliance of the Thai body politic under pressure of a putsch in May 1981 bodes well for the stability of the entire region. To construe these events as the collapse of a row of dominoes seems oversimple if not downright paranoid.

Antidomino

Since there is no formal statement of domino theory, in order to analyse its logical structure we can only examine the mechanics of the analogy. The elegant, rippling collapse of a row of dominoes derives from its artful arrangement in a state of unstable equilibrium so that any disturbance will be transmitted along the row. The pieces are endowed with potential energy by standing them on their ends so that each will strike the next as it falls. If a gap greater than the length of a piece separates two dominoes, the chain reaction ceases. The dominoes have three states: standing, falling and fallen. That 'falling' and 'fallen' equate with 'going communist' may satisfy the moral perspective of those who apply this theory. On the other hand they might have been disturbed that the fallen state was a stable equilibrium while standing was unstable. The red and white characterisation of politics implied by the analogy is not only naive and insulting but also runs contrary to a geographical sense of uniqueness. It utterly fails to capture the significance of regional or national identity which daily we see dominating mankind's sense of self and place.

The model treats of aggression from one end of the row as the potential energy of the first domino is translated to kinetic energy by an initial tap. It falls, registering a change to the same affiliation as the aggressor and, in so doing, imparts this character to the next domino as it strikes it down and so forth. What the necessities of similar size and appropriate spacing translate into in geographical terms is unclear. Obviously in order to land on the beaches of San Diego some very large dominoes would have to be stationed on the Philippines, Wake Island and Hawaii. The existence of a gap like the Pacific should quiet fears of the red menace wading ashore in California. In the proliferation of the theory's use, oceans or intervening nations are obviously not seen as gaps containing the contagion, but can be conveniently erased. The nature of the contamination process is not made very clear by the analogy. 'Knocked over' is redolent of liquor stores rather than nations and hardly provides a rich enough description of the process to prescribe preventative action. 'Propping up' has been used to indicate one type of solution, but has proven difficult to translate into successful political, military and economic operations. 'Knocking out', the lateral displacement of one or more pieces to provide a fire-break to check the progress of the conflagration, does appeal to some military minds as a feasible action. Certainly, on a local scale, towns and villages were wiped out to save them from the Viet Cong in Vietnam.

For the enlightenment of grand strategy it would seem more appropriate to conjure up a representation which is more explicit about the process of conversion and which identifies the epicentre of change more explicitly than a flick of the forefinger. As an ageless and general driving force behind the process of territorial competition among mankind we can establish the axiom that 'power ingests weaker centres of power or stimulates rival centres to strengthen themselves' (McNeill 1963). We can imagine a limitless plain upon which rival foci of power and identity may arise. For a given quantum of economic, military and moral power, the focus of an empire can only extend its field of influence at a cost in terms of the spatial intensity of this influence. The same total amount of power spread evenly over a wider radius will form a more squat cylinder than if it is confined to a narrower compass. The intensity of power exerted must decline as the inverse square of the radius of the territorial limit, giving a potential field of influence like a sagging bell tent. (This representation could be generalised to account for variations in distance friction and population density by defining a geographical coefficient which transforms the homogeneity of Euclidean space into the differentiated space of geography.) This abstraction would suggest that mustering sufficient power to overcome and ingest any erstwhile independent centres presents a greater strain the wider the bounds of an empire are cast. Independent foci require less energy to emerge within the bounds of a more widespread domain. The contest between two adjacent, compact empires is likely to be fiercer than that between two far-flung empires with distant cores. Competition between expanding empires will tend to generate an equilibrium boundary where their levels of influences are equal. There is ample historical evidence of the overextension of empires and their collapse as power is spread too thinly. More compact units such as Byzantium and China were far longer lived than the empires of Alexander, Rome, Ghenghis Khan, the Third Reich and the East Asian Co-Prosperity Sphere.

What has been proposed here as an alternative to the domino model is something akin to that employed for market area analysis. Similar structural and comparative static results to those derived for economic competition within a nation could be produced to investigate the nature of geopolitical competition. This would seem a richer source of conjectures and basis for refutation of fallacies than the elegant silliness of a row of dominoes. The results produced would, of course, be haunted by the same indeterminacies which dog the search for the certainty of equilibrium in the discussion of oligopolistic competition. But this would be a truer reflection of the realities than the simple symmetry of all fall down.

A first step in elaboration on this theme would be to relax the assumption of a fixed quantum of power spreading out from a singular focus. It could be contended that by gaining territory an empire gains a greater resource base and so its influential and material power expands with the radius of its frontiers at a rate of 2π in Euclidean space. Such an understanding, which must have fuelled the dreams of conquerors and empire builders, would seem over-optimistic from the historical evidence, certainly when it is taken to the limit of world domination. This simple geometry of territorial ingestion and expansion of a dominant influence limited only by the scope of the planet is at odds with evidence of the diffusion of cultural traits from the past which suggests a loss of expansive energy and deceleration of adoption as the limit expands. This would seem to underlie the S shaped curves that characterise the spread of innovations through time. In the last century several technical advances have spread universally but it still seems to be the case that the effort required to extend a cultural domain increases at the margin.

When we identify the cultural traits involved as political affiliation and command structure, then the components of increasing and decreasing returns to scale are more obvious. On the economic side it is nearly axiomatic that efficiency will increase with the scope of an integrated economy as barriers to exchange and specialisation according to comparative cost advantages are reduced. (These benefits can, of course, be enjoyed without territorial hegemony by a centralising power.) Counteracting this urge towards economic unification there seems to be a distinct political and social antipathy towards largeness. The need for membership of a comprehensible whole and the prospect of making an impact on the course of events, means that political effectiveness declines with the extent of the polity. It would be possible, if a little fatuous, to postulate the optimal size of the polity in terms of the balance of economic advantage in favour of bigness against political and social advantage in favour of smallness. Whether or not precise conclusions could be drawn, it is evident from the vehement affirmation of separateness of smaller groups within the confines of, or adjacent to the territories of world power cores, that the limited notion of nationality is still the most powerful geopolitical force. Those seeking *de facto* or *de jure* independence often turn to the chief competitor of their dominator for support. Cuba is the obvious example of this behaviour. An understanding of the relationship of North Vietnam and China in these terms might have been more enlightening to US foreign policy in the second half of this century than the domino theory.

A more felicitous mental picture of the potential for interaction of the three major power cores, their several more diffuse subsidiaries and the multiplicity of client, satellite, independent and refractory entities of international politics is conjured up by Henry Kissinger's term 'linkage'. Once again this verbal abstraction has not been formalised anywhere. The model behind this most characteristic device in Nixon-Kissinger policy is of a network connecting all the world's trouble spots to the Soviet Union and the USA. The resolution of particular conflicts then depends not on the merits of the case alone but on the overall balance of power between the two sides. It is clear from recent events that China was and is considered a card in this game rather than a player. This is obviously a silly misconception which China will manipulate to considerable advantage.

The geographic defect lay in the peculiar configuration of the network which Kissinger seems to have employed. The graph had two terminal nodes with all others connected to them by equal valued edges. Kissinger viewed the links from all US/Soviet points of confrontation throughout the globe as being of equal significance. In practice, the original linkage theory has been called 'little more than unreconstructed cold warriorism' and a formula for perpetuating confrontation all over the world (Landau 1972). This contrasts with a 'ripple' approach which sees events in the world only as strongly interconnected as the geographic or conceptual distances between them are brief.

A modified linkage model incorporating the friction of distance, with nations as nodes of a more fully connected network of economic and political links weighted in terms of the ease of communication and influence between states, would suggest the efficacy of indirect methods of achieving geopolitical goals. This would encourage the search for solutions by the diplomatic manipulation of second or greater order paths of influence, rather than head on military posturing or action.

Linkage theory is potentially a far more satisfactory model than domino theory, but its finite structure leaves much to be desired as a useful representation of geographical reality. Geography instills a predeliction for continuous formulations. The spaceless nature of the nodes of a network is likely to downgrade the significance of internal differentiation and politics in the workings of the state. Alistair Cooke remarked in one of his 'Letters from America' that you only had to spend a while with a map of Southeast Asia to dismiss domino theory. Although linkage gets you from the one dimension of domino theory into two dimensions, it still needs to be supplemented with a more elaborate intellectual picture of our world. Neither dominoes nor dyads will do as an adequate impression of geographic, cultural entities. Until such time as we shake off the coils of distance our motives and actions will be influenced by where we are and where everybody else is, and statesmen would do well to keep this in mind.

Chains and Arcs

Two similies which have been used recently to describe the geography of the confrontation between the USA and USSR present the danger of progress from a simplified mental map to a theory and a prescription for action. On the American offensive *The Economist* has written of China as an essential link in the 'global chain' Reagan wishes to tighten around Russia. The series of flashpoints and potential threats to Western European and American oil supplies which dot North Africa and southwest Asia from the Horn of Africa to Chittagong was labelled an 'arc of crisis' by National Security Advisor Brzezinski in 1978.

Perhaps the greatest disservice to peace that the chain image of containment does is to state baldly the reality of a recurrent Russian nightmare of being surrounded by a ring of powerful and coordinated enemies. It is possible, without straining the imagination too far, to interpret Soviet foreign policy as a thoroughly Russian quest for secure borders. The history of Russia certainly offers a ready explanation for this preoccupation. Any words or actions which amplify the Kremlin's perception of being encircled by aggressive opponents can only heighten fear and increase belligerence.

Consolidating a wide variety of political and violent conflicts over a swath of territory into an arc of crisis seems like the West feeding its own paranoia. The generalisation lends itself to a simple Kissinger linkage interpretation and response with Moscow's initiative seen at the back of every eruption, requiring a consistently pugnacious attitude regardless of the local circumstances. Such a response might, of course, not be consistent with the more mundane objective of securing oil supplies. An arc smacks too much of a battle front. To engage in diplomatic manoeuvres with the two-team game mentality of the battlefield is to limit seriously your own flexibility and the prospects of coming out ahead, while increasing the possibility of impasse and the resort to violence. An arc could too easily become a last ditch.

Communist Geopolitics

If geographic reality is misconceived by the opponents of communism, it is also the case that that world picture, seen from the camp which occasioned it, is subject to distortions in terms of received doctrine, its interpretation, and of geographical lore and prejudices inherited from the past. Russian, Chinese and Vietnamese, along with every other variety of politician and soldier are profoundly influenced by the myths and legends of experience stretching back far beyond Marx's revelation. The most oppressive danger to Mother Russia has always been seen to come from Germany. For Han Chinese it is part of the natural order to extend their harmonious ways southwards and to build a bulwark against the barbarians of the steppes to the north and west.

An examination of the gospel from which the doctrine of the avowedly communist state derives, reveals little to guide policy geographically. In this, as in many other things, *Das Kapital* got things wrong. Marx was impressed by the revolutionary barricades of Paris and the industrial growth of Great Britain. To him it was obvious that class war and the revolution would ignite in the urban west. There is little or nothing on the practicalities of class war and how and where it would be waged in Marx. Engels did observe the communards struggle in 1870 first-hand, and this experience contributed to his final scepticism about revolution. For the practice of revolutionary war the faithful must turn to Lenin for guidance. His pragmatism was not coupled with a great capacity for geographic synthesis. He expressed things in terms of unique, historical circumstances and political mechanics. The course of the civil war in Russia did not encourage any spatial generalisation. It was fought along class rather than territorial lines with the Bolsheviks and industrial workers fighting the former propertied and official classes for control over the wavering peasant majority. Lenin was un-sympathetic to any abstract formula, concentrating on the actual fighting and the political and economic circumstances which attended it. To be successful, it was necessary to create new methods to suit changing aspirations.

Finding themselves in charge of the anomaly of a successful commu-nist revolution in an essentially rural nation, which Marx had dismissed as the least likely participant, there was some confusion of aims and geopolitics among the early rulers of the Soviet Union. Trotsky saw the urban west as ready for revolution and proclaimed the duty of the Communist International to stimulate revolutionary activity there to weaken capitalist governments and prevent them from waging war against the Soviets. The Soviet Union would lead a permanent revolu-tion, bringing proletarian power to industrialised countries like Germany and Britain, who would then send manufactures to Russia in return for raw materials. Trotsky, Menev and Zinoviev made their bid for power on this basis in the summer of 1927. Stalin organised the isolationists' opposition on the platform of Lenin's policy of industrial-isation and collectivisation to achieve economic self-sufficiency. Trotsky and his illusions of global revolution were expelled in 1928 and in 1930 Stalin declared the defensive doctrine of capitalist encircle-ment. Litvinov crusaded in the 1930s for international cooperation to overcome the threat of German rearmament. In 1939 he was replaced by Molotov who vainly pursued a policy of conciliation with Germany and Japan to turn their attention elsewhere. By 1944, it struck close observers that the Soviet government had reverted to a pre-revolution-ary view of Russia's world interests and much of what she has done since is open to that interpretation. Stalin effectively nationalised revolutionary zeal.

Behind the pursuit of 'collective security' and the opportunist creation of a defensive wall of satellites in the aftermath of World War II, however, the Soviet elite retained a faith in the logic of history and the vanguard role of the USSR in fostering the inevitable collapse of the capitalist international order. With Khrushchev's assumption of power this spirit broke to the fore. Khrushchev believed that the USSR was on the threshold of achieving economic and military superiority. He repudiated the doctrine of encirclement with the observation that who was surrounding whom was open to question. Khrushchev thrust the USSR onto the global stage, promoting the socialist way among the tropical colonies from which Europe was in retreat. He also opened the rift with China, in the first place over revolutionary doctrine. This soon took on the shape of the traditional enmity between the two nations. Brezhnev has proceeded with less confidence in the inevitability of Soviet material success. Although the current leadership has continued to intervene to sway the course of events far from the Soviet borders in Africa, Asia and America to discomfort the capitalist order, it has limited action to relatively inconsequential places and sought stability in really dangerous settings like the Middle East.

The banner of world revolution was taken up by the successful heretics of the Chinese party who had mobilised the peasantry in violation of orthodoxy. Lenin, Trotsky and their successors were suspicious of guerrilla war and the lack of central control it entailed and here was a movement in which guerrillas played a central role in revolutionary development. It was Lin Piao who articulated a coherent geography of revolutionary war when he proclaimed the objective of encircling the cities of the world in 1965. This was Mao's theory of the rural base for revolution and urban encirclement writ big. Lin designated North America and Western Europe as the urban world and Asia, Africa and Latin America as the rural surrounds. Since revolution had been temporarily held back in the capitalist world, the cause of world revolution hinged upon the struggles of Asian, African and Latin American revolutionaries who could command the support of the majority of the world's population. It was clearly the duty of communist governments to support these usually peasant-based efforts. The geographic perception of this doctrine is every bit as preposterous as domino theory and as powerful in misguiding action and spawning fear. Che's unsuccessful 'foco insureccional' was a geopolitically conscious attempt to put this into practice. The Soviet Union remained wary of

guerrilla war and exerted considerable pressure on Cuba to stop finan-
cing rural movements in Latin America. In post-Mao China it seems
most likely that there will be an increasing emphasis on traditional
objectives in a Sino-centric world where significance diminishes radially
forming concentric zones around the eighteen provinces of China
proper.

Readings

A foundational work which provides a deeply geographical interpretation of
world history is:
W.H. McNeill's *The Rise of the West*, referred to in ch. 1

A collection of writings on world politics which traces the rise of American naval
power up to 1945 is:
H. and M. Sprout (eds.) *Foundations of National Power: Readings on World
Policies and American Security* (Princeton University Press, Princeton, 1945)

The seminal geopolitical work was:
H.J. Mackinder 'The Geographical Pivot of History' *Geographical Journal*, vol. 23,
no. 4, (1904)k, pp. 421-44

Geopolitics and Nazi policy was examined in:
D. Whittlesey *German Strategy of World Conquest* (Farrar and Rinehart, New
York, 1942)

The image of cores and shatterbelts is presented in:
S.B. Cohen *Geography and Politics in a Divided World* (Random House, New
York, 1963)

The domino theme is threaded through:
R.B. Asprey *War in the Shadows: The Guerrilla in History* (Doubleday, Garden
City, New York, 1975)

The picture of Han expansion is painted in:
H.J. Wiens *China's March Towards the Tropics* (The Shoe String Press, Hamden,
Connecticut, 1954)

The linkage model of geopolitics is considered in:
D. Landau *Kissinger: The Uses of Power* (Houghton Mifflin, Boston, 1972)
A.K. Henrikson 'The Moralist as Geopolitician' *The Fletcher Forum: A Journal
of Studies in International Affairs*, vol. 5, no. 2, (1981), pp. 391-414

Soviet attitudes up to 1944 are treated in:
S. Welles *The Time for Decision* (Harper and Bros, New York, 1944)

A more recent summary is to be found in:
The International Institute for Strategic Studies 'Prospects of Soviet Power in the
1980s' *Adelphi Papers*, Nos. 151 and 152 (1979)

Chinese attitudes are revealed in:

Lin Piao *Long Live the Victory of People's War* (Peking, 1965)

N. Ginsberg 'On the Chinese Perception of a World Order' in Tang Tsou (ed.) *China's Policies in Asia and America's Alternatives*, vol. 2, (University of Chicago Press, Chicago, 1968)

8 GUERRILLA WARFARE

Uproar in the east, strike in the west

Mao Tse Tung after Sun Tzu.

With the uneasy balance of mutual assured destruction holding the big battalions in check, most of the fighting being done is of the guerrilla variety. Indeed, since 1945 the chief type of war waged has been of the local, colonial or civil variety with irregular, guerrilla tactics employed by one or both sides. Much of the employment of regular forces has been in counter-guerrilla operations. Between 1945 and 1975 there were 54 colonial and civil wars with 8 million victims as opposed to 17 conventional, international wars causing 2.9 million deaths. If we measure the importance of wars in terms of the numbers killed in them, then since 1945 the small wars in the shadows have been of overwhelming significance. The potential flashpoints of global confrontation which might unleash the massive might of major conventional forces and trigger the horrifying prospect of nuclear exchange have been mostly limited to small fights involving irregulars. These fighters are not motivated by professional pride, *esprit de corps* or a sense of duty but equally or more potent sentiments such as religion, *machismo*, nationality or *compañerismo*. Guerrilla warfare has achieved considerable success as an auxiliary or prelude to more formal warfare or as propaganda by action to exploit political weakness, create a power vacuum and an elite cadre to fill it. Lawrence, Collins, Begin, Tito, Mao, Giap, Grivas and Castro, among others, employed these tactics to the considerable benefit of themselves or their political heirs. The mobile tactics of *guerrilleros* have been adapted by regular units. Units were formed and named for this particular function, as in the commandos, raiders and rangers. Liddell Hart found an inspiration for his advocacy of indirect and mobile war in the exploits of T.E. Lawrence. The use of Orde Wingate and his Chindits, the Long Range Desert Group, the Commandos, the Green Berets and the SAS (Special Air Services) is in the long tradition of employing mobile partisan units in disruptive fashion, as Kutuzov turned Davydov and his cossacks loose on Napoleon.

The Time and Place for Guerrilla War

In the latter half of the 20th century the territorial competition between ideologies and empires is being fought out in petty wars in the shatterbelts. Through the centuries these wars have flared or smouldered around the globe attracting attention in proportion to their impact on imperial pretensions. Since the tactics of guerrilla war, such as hit and run, the lure and ambush, are doing what comes naturally in fighting, their use is as old as political bellicosity. Joshua used them, as did Scythians, Parthians, Fabius, Vercingetorix, Goths, Huns, Vikings, Turks and Mongols. British and French colonists relearned them from native Americans and turned them on each other. Rogers' Scouts and Marion's South Carolinians provided Anglo-Saxon America with early exemplars.

The tactics have been used both by those who are trying to upset the way things are and by those trying to preserve it. It was antipathy to France's revolutionary radicalism, however, which provided three of the classical instances of such war. The bocage of western France sustained the Vendée; the sierras sheltered the Spaniards who gave the genre its name; while the Russian army used partisans and cossacks to harry the Grand Armée in 1812.

The colonial conquests of the 19th century brought imperial armies into conflict with irregular formations of tribally organised cultivators, hunters or pastoralists. The Russians were opposed by Shamyl in the Caucasus and Poles in the west; the French faced Abd el Kadar in Algeria and Suarez in Mexico; the United States army was mauled by Morgan and Mosby in the South and Sitting Bull and Geronimo in the West; the British battled Zulus and Maoris, Ashanti, Boxers and Boers. The Spanish empire, ebbing before the rest, left a heritage of political instability, guerrilla wars and their patron saints such as Martí, Sandino, Zapata and Villa in its wake.

The action around the General Post Office in Dublin in 1916 began the transition from the tradition of the barricades to the era of the 20th century urban guerrilla. The following year saw revolution in Russia and a flurry of guerrilla action by the Bolsheviks and their opponents. The dangerous example of the White guerrilla groups led Lenin and Trotsky to close down Red guerrilla operations in favour of more readily controlled regular forces as soon as possible. The Italian, German and Japanese bids for world empires in the 1930s and 1940s generated partisan opposition, frequently fostered, trained and supplied by the Allied war apparatus. Although their military impact on the outcome of World War II was slight, the guerrilla organisations became political forces in the fracture zones of the post-war world, winning power rapidly in China, North Vietnam and Yugoslavia. The circumstances which left the Kuomintang battered, and pressed Japan to withdraw from China to meet the challenge in the Pacific, enabled the communists to win power. Much of the credit for this success was then ascribed to guerrilla warfare as promulgated by Mao. His utterances on this subject were mostly lifted from the brilliant 2,400 year old commonsense of Sun Tzu. No matter the source of Mao's wisdom and the reality of events, the myth of the military potency of rural guerrillas gave credence to Lin Piao's 1965 exhortation to the world's peasantry to engulf the urban-industrial world.

The continued extension of Russian, Chinese and American influence of the second half of this century and the implosions of Western Europe's political realm provided the setting for a succession of little wars. Sinn Fein and the IRA waged a partly successful campaign of disruption and terror from 1919 to 1922 and the Zionists followed their example in 1944 to 1948, providing an object lesson for the future PLO. In Vietnam, Indonesia, Malaya, Kenya, Algeria, Cyprus, the Congo, Angola and Mozambique, colonial powers were attacked with varying degrees of military success. The cost, irritation and brutality involved seem to have precipitated what would have been inevitable withdrawal. The turbulent politics and inequitable distribution of wealth through much of Latin America have always provided opportunities for daring power-seekers who are willing to incite bloodshed. One such escapade, by Castro in Cuba, succeeded with little military exertion because of the general antipathy to the regime of the incumbent, Batista, and its corrupt fragility.

In Indochina the anticolonial battle of the Vietminh became expansionary, pushing out the French and then pressing the Diem government. The USA got embroiled on the basis of the geopolitical misperceptions incorporated in the domino theory. It was the panic over the fall of Dien Bien Phu which brought the attitude to a head. This was not, however, occasioned by guerrilla warfare. General Giap's main manoeuvres at this time were of regular formations. When the US came to be directly involved, the marines had some basis for claiming to have won the shooting war with the Viet Cong guerrillas. But the war of nerves, opinion and politics was won by the *calico noir* and US forces extricated themselves painfully.

In the circumstances of our increasingly urban, mobile and interconnected world, the chances of success for battalions of irregulars moving around in more or less permanent battle order, faced with a moderately competent incumbent, have dwindled. In predominantly urban societies with a small contented rural population but with city dwellers suffering the dislocations of rapid change, guerrilla formations have in places fragmented into terrorist bands, more in keeping with the atomised structure of society and the potential for propaganda of the news media of the global village. Northern Ireland stands in an odd congealed transition between the tribal loyalties of rural society and urban fragmentation. In 1969, the smouldering antagonism between protestant and catholic populations was fanned into a shooting war between their zealot fringes. With this the dormant IRA got an infusion of fresh blood, separating into radical Officials and the Provisionals with more local loyalties. The near success of the FALN in Curacao in 1965, the failure of the Guevara to generate rural revolutionary foci in the Bolivian chaco, and the preaching of Guillen and Marighella ushered in an era of urban terrorism in Latin America. This was made manifest by the ALN in Rio de Janeiro and São Paulo, the Tupamaros in Montevideo and the Monteneros in Buenos Aires. Where the military and police opposition is strong, where popular support is limited and where motives are predominantly destructive, guerrilla action has degenerated into terrorism. In the 3rd century BC Wu Ch'i pointed out that one man willing to throw away his life is enough to terrorise a thousand. Television, radio and newspapers amplify this several-thousand-fold. In the meantime the example of Iran in 1978 showed that under auspicious conditions power can still be seized by irregular forces. The failure of the concerted effort of guerrillas in El Salvador early in 1981 suggested that success does not come easily. The events of early 1982 indicated that it is possible to shake the regular military severely while the election of April 1982 called the popularity of such movements into question. In Afghanistan there is evidence that heading for the hills is still a viable military proposition. In Zimbabwe the battles between Shona ZANLA and Ndebele ZIPRA point to *patria chica* as the most powerful motive among guerrilla fighters. The successful South African strike against SWAPO targets in Angola in August 1981, illustrated the potency of well prepared and equipped regulars against guerrillas.

The Significance of Geography

Theorists and commentators on guerrilla fighting all stress the importance of geographical factors in determining the outcome. It involves an intense exploitation of the character of the landscape. At the practical level, geographical skill has been deemed essential for successful operations. In 1759 de Jeney prescribed that the partisan leader 'keep some geographer under his orders who can draw up correct plans of the armies' routes, their camps and all places to be reconnoitred'. In 1970, Marighella was admonishing urban guerrillas 'to have knowledge of topographical information, to be able to locate one's position by instruments or other available resources, to calculate distances, make maps and plans, draw to scale . . .'

It is evident that any analytic understanding of this form of war must be infused with a highly developed sense of geography and feel for the landscape. China's greatest military geographer Ku Tsu-Yu wrote in 1698 that 'no one can discuss strategy better than Sun Tzu, and no one can discuss the advantages of terrain better than he'. In his treatment of guerrilla warfare von Clausewitz cited five general conditions for its successful pursuit:

1. that the war is carried on in the interior of the country
2. that the outcome cannot hinge on a single battle
3. that the theatre of war extends over a wide area
4. that the national character is favourable to war
5. that the country is irregular and difficult, being mountainous, wooded or swampy or because of the kind of agriculture.

All but the second of these are geographical conditions.

Guerrilla and Counterinsurgency Tactics and Strategies

There are no clear-cut lines dividing the continuum that runs from regular formations through to fragmentation into individual terrorists in describing the political use of violence. There is, however, a generally recognised intermediate variety of fighting where irregulars or partisans combine military and psychological objectives in their operations. By contrast with conventional fighting, there are two distinct viewpoints from which guerrilla warfare is viewed. The options and opportunities for guerrilla action are quite distinct from those which confront the opposition, counter-insurgency operations.

The literature on guerrilla tactics is extensive, much of it wisdom dispensed by former practitioners who wield the pen as easily as the sword. Most of this is elaborated common sense and reinvention of the wheel. History is replete with leaders who never had the benefit of these handbooks and did the right thing by instinct or learned rapidly from experience. The essence of guerrilla tactics is to trade space for time. The enemy is allowed to dominate a lot of territory and his morale and force is slowly eroded by a thousand small cuts. He is drawn to extend his supply and communication lines and spread his firepower thinly so that his internal connections as well as his flanks may be gnawed and his resolve eroded by constant nipping. Hit and run raids, diversions, sabotage, terrorism and ambush are the principal kinds of engagement. Although feinting and running to avoid pitched battle are primary ploys, and strategically guerrilla war is usually defensive, to achieve success it must be tactically on the offensive. Obviously, it falls into Liddell Hart's category of indirect methods and guerrilla action provided one of the inspirations for his advocacy of movement and surprise. In effect, he turned the offensive-defence of traditional guerrilla war on its head to produce the defensive-offence of deep penetration, in which the perfection of strategy would be to produce a decision without any fighting. His dispersed strategic advance avoided the concentration of force, maximised the enemy's uncertainty and dislocated his interior lines, organisation and mental equilibrium. The mobile and deceptive tactics required for both guerrilla fighting and Liddell Hart's variety are best summarised in the slogan which Mao adopted from Sun Tzu and with which we began this chapter. The thought which he paraphrased most frequently is verse 12 of chapter 7, 'Now war is based on deception. Move when it is advantageous and create changes in the situation by dispersal and concentration of forces'.

The ultimate objective of both sides in guerrilla war is control of the people. Thus, in order to succeed any force countering guerrilla action must not only defeat guerrilla forces militarily, but also attain the political goals of separating the population's sympathy from the guerillas and ensuring the existence of an acceptable social order and government. A campaign to 'search and destroy' must be combined with a reasonable prospect of winning 'hearts and minds' and sustaining a viable political authority. The tactics employed will depend on how embedded in the terrain and the social fabric the guerrillas are. To face the adventurous incursion of a small group in an alien setting, such as Guevara's Bolivian effort, a mobile strike may be sufficient. In the case of an indigenous movement with some popular following, such as the Viet Cong, flying columns have to be combined with an effort to control territory. 'Hammer and anvil' exercises and aggressive patrolling needs to be coordinated with 'cordon and sweep' or 'quadrillage'. Some combination of territorial control and mobile striking power is usually called for.

The conservatism of warriors is legendary and much of the difficulty regular forces have had with guerrillas has arisen from their inability to scale down the scope of their operations to match the 'intangible web' of the opposition. One remarkable historical example of rapid and successful adjustment to a physical and human terrain conducive to defensive, guerrilla warfare is the Norman conquest of Ireland after 1169. Eschewing the feudal trend towards the full frontal charge of heavy cavalry, Strongbow's men routed Gaelic swordsmen with their mobility through bog and forest and their cunning deviousness in seeking battle. Having won the ground they held it by encastellation, the building of small strongholds at controlling points in the landscape. In terms of winning the sympathy of the population, they very rapidly went native, adopting the language and style of the aristocracy they displaced. Their success had the two essential elements of a mobile and flexible strike force and an effective means of holding what it won in the long term. What they devised as a strategy immediately has had to be relearnt painfully time and time again.

Taking the fighting component first, the main need is to get just sufficient steel or firepower to overwhelm guerrilla bands wherever they can be found, at rest or in action. This implies small, fast moving units — the flying column. French colonial experience in Senegal, Sudan, Indochina, Madagascar and Algeria led Gallieni and Lyautey to elaborate the 'tache d'huile' method of countering insurgents. The military element of this, which preceded the civil operations, depended on achieving the greatest speed and most intense intelligence possible. An initial target region was partitioned into operational areas. This was 'quadrillage'. In the next step, 'ratissage', the force assigned to a quad-rille cleared out insurgents with fast, ever-moving units seeking friendly ties and intelligence from the population. Their speedy, swirling movements were described as 'tourbillon' — whirlwind tactics.

In the 1950s, faced with more severe opposition in Indochina, Salan adopted a version of Orde Wingate's 'stronghold' concept with his 'bases aero-terrestres'. These were air-supplied garrisons placed in remote regions where the insurgents had the upper hand. The object was partly flag showing but also to draw guerrillas into the open where firepower superiority would tell. If the guerrilla forces can muster enough firepower of their own, and where the base is poorly located in a saucer, as in Dien Bien Phu, this is unlikely to be successful. The USA repeated the experiment in Khe Sanh in the late 1960s. The claim of victory there rings hollow, since Giap merely encircled it, sucked in marines and held the upper hand in the region while he built up for the Tet offensive.

The chief problem which regular forces face in dealing with guerrillas is the need to scale down their operations geographically and numerically. This is especially so for armies which have been victorious in massive conventional conflicts. The British in Malaya had the advantage of a cadre who had been with Orde Wingate's Chindits in Burma, learning to deal with rainforest and jungle clad hills and to operate in small, independent formations. The man in charge of affairs, Templer, was an astute politician who addressed the circumstances as a police problem stressing the maintenance of security at the village level and the ultimate goal of an independent Malayan state. In Cyprus, Harding does not seem to have been so politic and, according to his adversary Grivas, supersaturated the island with troops who provided targets and got in each other's way, so as to actually diminish the degree of control achieved. Grivas suggested that he would have done better with small units of specially and intensely trained anti-insurgents. The British had done this successfully in Malaya, resurrecting the SAS and employing Royal Marines in this capacity. Experience and methods devised in Malaya were transferred to Kenya, Aden and Borneo. The SAS were dedicated to this special function and have more recently been used in Northern Ireland and against terrorist activities in England, most notably in the case of the Iranian Embassy's seizure. In 1961, when there was a flurry of political interest and concern over guerrilla wars, the US established the Green Berets, geared to win hearts and minds and provide deep focuses from which to defeat guerrillas.

When the USA became directly and totally involved in Vietnam in 1965, their theatre commander was Westmoreland, an artillery man with a quantitative bent. The first major ground forces to wade ashore were the marines. They simply adapted their amphibious mode of operation to helicopters in a tactic of vertical envelopment. The airborne cavalry's operations were essentially of the same nature. The emphasis in operations was on winning the war as measured on some objective scale. The number of people killed, euphemised as the 'body count', became the objective of the exercise. This was pursued with search and destroy spoiling operations, seeking to block and envelop the Viet Cong. The emphasis was on killing rather than holding territory. The simple military goal of killing was grasped and it overwhelmed the ultimate objective of pacification, whose nature and measure of achievement was less tangible. The attitude is summarised in the words of a senior US officer, 'Grab 'em by the balls and the hearts and minds will follow'. When Abrams took over command in 1968, he did begin to resolve operations down to a more appropriate scale for the environmental and political setting with 'stingray' tactics. These involved smaller detachments sweeping areas more frequently and faster.

The holding and pacification functions were an integral part of Lyautey's tache d'huile strategy. The military actions were only the necessary prerequisite and were alone insufficient to attain the desired end state. Once cleared of insurgents the quadrille was taken over by the civil authorities who won over the tribesmen with protection and the prospect of economic and social progress. The oil dropped on these spots spread French influence slowly and methodically, over the land. The British in South Africa, harried by the superior tactics of the Boers, used a more brutal method of controlling the civilian population which was providing logistic support to the Afrikaner commandos. The Boer's scattered farms were razed and the non-combatant population was gathered in concentration camps. The methods used in Ireland to pick through the civilian population looking for guerrillas, evolved in Palestine into cordon and sweep operations, did little to endear troops to the populace and, indeed, helped guerrilla recruitment.

British success in Malaya was founded on the reform of land tenure which established the titles of Chinese squatters; on the 'new village' programme which increased the security and ease of policing the rural population and, finally, the clear commitment of the authorities to the creation of a self-governing, multi-racial society. In 1959, the Diem government in South Vietnam started a similar 'agroville' programme to fortify hamlets and villages against Vietminh attacks. In 1962, on advice from Sir Robert Thompson based on his Malayan experience, a programme was mounted to develop a network of 'strategic hamlets', combining military and political purposes. In practice this provided the Diem government with an excuse to spend American aid money to increase their political surveillance of the population. Instead of a steady, oil-stain like spread from the Mekong delta as Thompson suggested, the net was cast over the entire country and over-extended. An effort was made to fortify 11,000 hamlets (two-thirds of the total) before the end in 1963. The civil and military personnel and apparatus to operate this sytem did not exist. The money involved provided a honey-pot for corruption and in many instances the hamlets were little more than concentration camps. What was lacking more than anything was the personally intense relationships which the Viet Cong cultivated with the people.

A number of French officers had come away from Indochina in 1954 convinced that the power of guerrilla warfare lay in its Marxist ideological inspiration. The theory of 'la guerre révolutionnaire' called for the development of a strong counter ideology with which to indoctrinate soldiers and civilians for the fight against Marxism. The strong fascist overtones of this doctrine as it emerged did not prove very popular at the time. The variety of guerrilla motives and settings through history give the lie to the simple and singular importance attached to Marxism. Nationalism remains the single most potent force in stirring guerrilla or conventional warfare.

A Geographical Analysis

The political and psychological dimensions of the balance of advantage between guerrillas and counter-insurgents is possibly amenable to a geographical treatment, but this would imply the general acceptance of a theory of history which we have not yet achieved. Until such time as this emerges there are two kinds of insight which a geographer can provide. Firstly, it is possible to provide a description and analysis of the unique circumstances surrounding particular events. Laqueur, the doyen of studies of guerrilla and terrorist activities, is inclined to credit this specific approach as the most valid one. There is a large volume of literature of this kind. Much of it would have benefited from the attention of a geographically trained perception. What is offered here, by contrast, takes the more abstract approach of generalising on the geographical settings in which guerrilla warfare may be successful, or otherwise.

The measurement upon which the debate on the efficacy of guerrilla or counter-insurgency operations turns is the ratio of regulars to guerrillas. The conventional wisdom of military and media lore is that it is necessary for regular forces to outnumber insurgents by at least ten to one in order to stand a chance of defeating them. This has been repeated recently in the context of El Salvador. The ratio appears in the literature on guerrilla war but there is no analytic basis supporting the contention anywhere. Nevertheless, it is clear from historical evidence that guerrilla fighting is highly effective in manpower terms. It has most usually taken a large number to even contain them. In the Spanish campaign of 1809-12, 3,000 guerrillas kept 18,000 Napoleonic troops busy, giving a ratio of six to one. The British needed a superiority of twenty to one to defeat the Boers. In Algiers, in the 1950s, the French were defeated in spite of having a similar predominance. In the Troubles of 1919-21 the 3,000 strong IRA successfully engaged 43,000 police and troops. (All of the above numbers are from Laqueur 1976.) The 10:1 ratio gained wide currency from the pronouncements of Generals Maxwell Taylor and Westmoreland when commanding in Vietnam. In 1965, however, Taylor proclaimed that a superiority of 25:1 would be necessary to defeat the Viet Cong. This was in recognition of the fact that given their vast support needs and low 'teeth to tail' ratio, the US and South Vietnam could only muster a superiority of 5:1 in fighting men at that time.

To investigate the geographical dimension of this measure we can only turn to the historical record. This is rather spotty. Table 8.1 contains numbers of guerrillas and their opponents from a selection of wars since 1945. These data are drawn from a variety of sources and are all subject to large possible errors. These errors arise from a lack of information or from the exercise of political deceit. Given the spontaneity and loose organisation of some guerrilla campaigns and the ease with which combatants can melt into the civilian population, there is always a difficulty of counting heads and allowing for the large fluctuation of numbers from time to time. In some cases numbers are exaggerated or understated for propaganda purposes by either party to the conflict. At the time of writing there is a heated debate generated by Mike Wallace of CBS accusing General Westmoreland of suppressing CIA data on the Viet Cong. The implication of the accusation is that a realistic estimate of 600,000 as compared to 285,000 might have caused President Johnson to withdraw in 1968.

When the guerrilla organisation extends across political frontiers there is a problem of determining how many fighters were operating across the border. Even within a state it is difficult to decide who is actively engaged on the guerrilla side in providing support. Similarly, with the incumbent forces it is not easy to determine what part is engaged solely in anti-guerrilla operations; the degree to which police and militia are involved in these; what the proportion of combat and support personnel is and what part of the support effort is dedicated to anti-guerrilla operations. The existence of auxiliaries or clandestine support for the status quo, such as the OAS in Algeria or the Ulster Volunteer Force, further complicates the issue of numbers.

To meet these problems, observations have been limited to cases where the main function of the military in a country was to counter insurgency. Figures for all regulars, police and militia engaged have

Table 8.1: Guerrilla War Combatants and Population Characteristics

Wars and Winners (G = Guerrilla Win, R = Regular Win)	Number of Regulars[1]	Number of Guerrillas[1]	Number of Guerrilla[2]	Regular force to space ratio (combatants[2] per square mile)	Guerrilla force to space ratio (combatants per square mile)[2]	Population density (combatants per square mile)[2]	Regulars per Guerrilla
Malaya (R) 1948-52	350,000	12,000	558	6.90	0.23	13.2	29.2
Cyprus (R) 1955-59	20,000	970	525	5.59	0.23	113	20.6
Algeria (G) 1954-61	560,000	30,000	318	0.61	0.03	10	18.6
Cuba (G) 1957-59	30,000	2,000	2,916	0.67	0.04	13.2	15.0
Namibia (R) 1966-	45,000	3,700	230	0.14	0.01	3	12.2
Uruguay (R) 1969-72	34,000	3,000	949	0.47	0.04	39	11.3
Greece (R) 1944-45	250,000	23,000	339	4.94	0.45	154	10.9
Ethiopia (R) 1960-	192,000	25,900	1,147	0.42	0.05	65	7.4
Guatemala 1981	20,960	3,000	2,420	0.49	0.07	173	7.0
Kenya (R) 1952-56	56,000	12,000	450	0.24	0.05	24	4.7
El Salvador 1981-	22,000	6,000	710	2.66	0.72	516	3.7
Palestine (G) 1945-48	100,000	54,000	3.2	12.16	6.57	167	1.9
Indochina I (G) 1945-54	342,000	200,000	114	2.70	1.58	180	1.7
Indochina II (G) 1960-73	941,750	600,000	70	7.40	4.70	331	1.6
Afghanistan 1981-	125,000	90,000	194	0.48	0.34	67	1.4
Cambodia 1977-	34,000	200,000	39	0.18	1.80	43	0.2

Sources: 1. Numbers of combatants were taken from:

a) Malaya and Algeria from A. Campbell *Guerrillas: A History and Analysis*, (Doubleday, New York, 1980)

b) El Salvador, Guatemala and Cambodia from *Time*, 7 September 1981 and 25 January 1980.

c) Palestine, Cyprus, Kenya and Indochina I and II from M. Carver *War Since 1945*, (Wiedenfeld and Nicolson, London, 1980).

d) Cuba and Greece from R. Asprey *War in the Shadows*, (Doubleday, New York, 1975).

e) Namibia, Afghanistan and Ethiopia from *Strategic Survey 1976-77* and *1980-81* (International Institute for Strategic Studies, London).

f) Uruguay from R. Moss *Urban Guerrillas: The New Face of Political Violence* (Temple Smith, London, 1972).

2. Populations and areas were taken from the appropriate year's issue of *The World Almanac*, (New York).

been used where appropriate, but unofficial participants have been excluded. Figures for all the guerrilla 'effectives' involved, whether their base was within the state or not, were used, although effectiveness has no concise definition.

Clearly, there are some conflicts labelled guerrilla wars which are no more than irritations to the established order and they hold little prospect of victory, military or political. Such engagements must be excluded from consideration since the guerrillas are hardly effective in any real sense. One possibility for establishing effectiveness is to use only the numbers of the victors in past contests in trying to discern the relationship between the number of combatants, area, density and terrain. This excludes present struggles, halves the rest of the observations and raises the problem of defining victory. The outcome of guerrilla campaigns may not be settled militarily but politically and the effort may bear fruit long after the active fighting. External circumstances weigh heavily on many outcomes. Frequently guerrilla activity has merely preceded or accompanied an inevitable withdrawal from global empire or change of regime. There is, thus, a justification for tabulating the numbers from both sides of the conflict for the time at which both were most active and the field of battle was most hotly contested. At this juncture both sides could be said to be at their most effective in military terms, reaching a temporary equilibrium of force.

Only wars since 1945 were used to ensure a degree of technological constancy among the cases, so that the effects of population density and terrain would not be confused by radical differences in weapons, vehicles and communications. We can assume that regular forces at least had air support, helicopters and mechanised transport and that the potency of weapons was much the same over this span.

When it comes to standardising the numbers for this analysis, the problem of delimiting the theatre of war arises. To relate variations in the ratio of regulars to guerrillas to geographical variables it is obviously necessary to draw a boundary around the action. Since guerrilla war thrives on fluidity and best avoids a territorial base, this is difficult. The areas of operation shaded on published maps are seldom clearly bounded. For the sake of consistency, and in the absence of better information, the area of the whole political unit involved is used here, seeking cases where this was more or less the fighting ground and avoiding those where this was obviously not the case. This does rather exclude the recent, solely urban campaigns. In other instances where the assumption departs from reality, this should show up as a departure from any discernible trend.

The data do confirm the economy of guerrilla fighting, since Cambodia is the only instance where they outnumbered regular forces. Even here the rebel side was swollen by Vietnamese assistance. It has taken a superiority of greater than 10:1 to win in four instances out of seven. 10:1 sufficed in Greece but less were needed in Ethiopia and Kenya. There seems no simple relationship between this ratio and the degree of sympathy for the rebel cause. As a proxy for involvement column 3 lists the population per guerrilla. The obvious geographical variables which might come into play in determining the effectiveness of the two sides in guerrilla wars are the availability of targets and support and the texture of the landscape offering cover and conceal- ment. On the face of it, accessibility, the ease of movement over the land, would seem an important consideration. A little thought, how- ever, suggests that this quality is neutral as between the two sides, if we presume an ability to learn and no long lasting technological gap between them. Guerrilla horsemen only survived on the plains until the light cavalry flying column equalised velocity. The cossack success against Napoleon was achieved on the forest fringe not on the open steppe. Helicopters and strike aircraft reduce the advantage of greater mobility of guerrillas over regular formations. Lawrence's operations in sandy desert would not last long today. In an urban setting, if both sides have cars, the effect of accessibility is neutralised. The availability of cover to fade into is the principle guerrilla requirement and even this is being penetrated by infra-red detection.

If we consider the density of support and targets variable first, the number of guerrillas needed to dominate combat in a given area should increase with the number of targets and the complexity of the human landscape to be controlled. The population provides at the same time potential material and moral support for guerrilla campaigns and also the density of targets for a war which must register as a constant irri- tant to the *de jure* authority. It may be that at very high densities, with strong local support, the high level of civilian logistic support and intelligence provided would lead to a downturn in numbers needed to dominate space. The number of regular, counter-insurgency troops required will also increase as population and target density increases, but at a faster rate since there is a need for comprehensive control rather than the sporadic exploitation of advantage of guerrilla tactics. The example of Cyprus suggests that at very high densities saturation may occur and potency falls off with increasing numbers as soldiers start getting in each other's way. The ratio of regulars to rebels neces- sary for one side or the other to succeed will be governed in its variation with population density by the scalar effect of the degree of sympathy for the guerrilla cause. The greater the popular support, the more effective a guerrilla, and the larger the number of regulars needed to counter a guerrilla at each density level.

Turning to the cover variable, obviously the number of guerrilla fighters needed to dominate a given amount of space will decrease as the amount of cover available increases. Thus, guerrilla effectiveness increases with cover. By contrast, the more broken and accidented the landscape, or the more textured it is, the larger the number of regulars needed to comb it and establish command and so counter-insurgency effectiveness could be expected to decrease with cover. The shift of insurrectionary movements into the cities of the world in the last decade takes advantage of the intensity of the urban fabric. The similarity of the urban landscape to naturally broken terrain was pointed out by James Connolly in 1915, likening the streets of a city to a mass of glens and passes. Connolly did not, however, learn in time to fight the fluid war that this invited.

In the course of the discussion we have introduced Liddell Hart's measurement of military intensity — the 'force to space' ratio. This was calculated for each side in each war and tabulated as columns 4 and 5 of Table 8.1. The density of targets and support is given by the population densities in column 6. It is obviously not easy to quantify terrain texture and not reasonable to apply a single measure to a whole national territory. To a degree, texture will vary in concordance with population density, and so we may in this preliminary analysis use this as a surrogate for both of our geographical variables. To examine how the effective force to space ratio varies with density we plot values for winners only, either guerrillas or regulars, in Figure 8.1.

The points plotted are too few and particular to constitute observations of a random variable amendable to probabilistic treatment. There is, however, a discernible tendency which is indicated by the trend lines sketched in for regular and guerrilla successes. It seems clear that the generally greater effectiveness of guerrilla manpower becomes more potent at higher densities where targets, support and cover are potentially greater. In order to succeed in high density, intense cover conditions, government forces must go far beyond the 10:1 mark. In sparsely populated, open country, they can succeed with less than this superiority in numbers. This is a clear invitation to bring guerrilla warfare into the cities as the Tupamaros did. An extrapolation of this divergence of effectiveness goes a long way to explaining the shift of insurrectionary emphasis to towns.

Figure 8.1: Variation in the Force to Area Ratio with Population Density

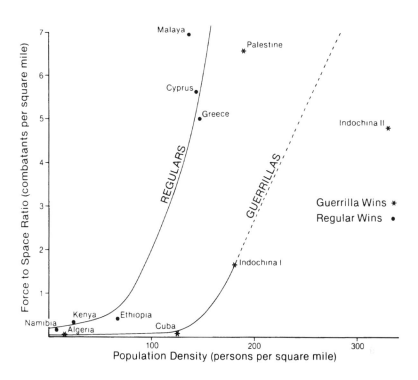

In all of this what tips the balance of numbers is the will and ruth-lessness of the protagonists. In China, Japan easily contained the 400,000 Red guerrillas with 250,000. On the other hand a handful of people employing terror can cause a lot of grief. The Symbionese Liberation Army was ten people and Britain's Angry Brigade was eight strong. The balance in these matters is more than purely one of destructive power. There is a current of opinion that the British Army could eradicate the IRA if they acted as ruthlessly as the Soviet Army using the tactics devised by Tukhachevski in the 1920s to deal with the Whites. In 1916 the British Army learned the lesson that if they 'put in the boot', in Ireland at least, the level of popular support for 'the lads' increases accordingly. Stepping up the intensity and violence of operations recruits directly for the opposition. Soviet intervention in Eastern Europe might well find itself generating such an effect. Even in Afghanistan the Soviet Army must be wary that the all-out employment of their 90,000 troops and mass of hardware to extirpate the rebels no matter the attrition of the civilian population is likely to reverberate in their own Islamic borderlands, where there are living memories of the Basmatchi movement which survived among the Turkic peoples of the Afghanistan border area into the 1930s.

Clearly the external circumstances, local details and chance which come into play in determining the final outcome of violent conflict make any attempt at formulation and quantification highly conjectural. Nevertheless, a practitioner of guerrilla warfare as perceptive as T.E. Lawrence did hold that 'the algebraical factors are in the end decisive', meaning by this, factors calculable in similar terms to those presented here. It is, thus, worthwhile pursuing the issue along these lines to discover the boundary conditions on the balance of superiority at least.

Readings

Asprey's *War in the Shadows* referred to in the last chapter is a masterful history of guerrilla warfare leading up to an analysis of US involvement in Vietnam. Other general works include:

R. Clutterbuck *Guerrillas and Terrorists* (Faber and Faber, London, 1977)
W. Laqueur *Guerrilla: A Historical and Critical Study* (Little Brown, Boston, 1976)
Mao Tse Tung *Guerrilla Warfare* introduced and translated by S.B. Griffith (Praeger, New York, 1970)
R. Moss *Urban Guerrillas: The New Face of Political Violence* (Temple Smith, London, 1972)

We have drawn freely on writings of practitioners and commentators put together as:
W. Laqueur (ed.) *The Guerrilla Reader: A Historical Anthology* (Meridian, New York, 1977)

9 URBAN WARFARE

> Attack cities only when there is no alternative
>
> Sun Tzu, (chapter 5, verse 7).

Towns focus wealth and power in the landscape. In this they invite the attention of conquerors or spoilers. Some large settlements existed for defensive purposes. In regions subject to the depredations of parasitic warriors or political enemies, farmers have gathered their houses together seeking safety in numbers at a considerable cost in terms of travel to and from their fields. The walled villages of Iran, the Berber fortresses of the Atlas and the kibbutzim of Israel are settlements of this kind. These are not, however, urban in the sense of housing non-agricultural activities such as commerce and manufacturing. The town is in essence a transport phenomenon. Buildings housing people, machinery, goods, services and information are drawn together to reduce the friction of distance entailed in production, exchange, governance and leisure. The advantages of close proximity are traded off against the need for space involved in various activities. Territory can best be controlled politically and economically by holding its urban foci. Towns then are the primary targets of war. In societies where over eighty per cent of the population dwells in urban areas, to seize control a would-be conqueror must fight for the towns. Even in predominantly rural societies, towns are of overwhelming importance in that they articulate the supplies of material and the power of command. Given their military significance, the location and layout of towns have often been selected with defence in mind. To dominate the land militarily and administratively and to oversee the economy, towns were built at key places on natural or man-made routeways. At the more tactical level, the sites for towns were frequently established to hinder the attack and aid the defence. Toledo, on a near circular bend in the Tagus, and Durham in a similar curve on the Wear, are beautiful examples of commanding settings and sites. In the case of Toledo, the town stands at the head of the incised gorge of the river, controlling the last crossing-point before the gorge and provides an obvious base from which to sally out to cut off attempts to cross further upsteam. Durham stands on the main passage along the eastern flank of the Pennines.

The architectural answer to the attacks of men, horses and missiles was to build a wall. To reduce the cost of this it was desirable to limit the area to be contained and this dictates a dense and compact arrangement of buildings. Although the street layout of the urban design which derives from the camps of charioteers has a square grid providing for easy passage, the essentially defensive objectives of medieval cities dictated an offsetting of street corners to act as a baffle against heavily armoured intruders, forcing them to turn frequently and concealing what lay in their path. Gunpowder put an end to the effectiveness of walls. To withstand bombardment or undermining some defensive depth was required. Arrangements of ditches and ramparts with bastions overlooking the field before them and covering each other were thrown around Italian cities in the 16th century. The science of such defenses reached its highest expression in the works of Vauban, around Lille, for example, dominating the passage to Flanders. Once the walls or earth works of a city were breached, however, since the besieging numbers were usually superior, to continue to fight through the streets was to invite certain slaughter and surrender held out the only real hope of survival.

Within the city, the narrowness of streets and mass of buildings provided good ground for the lower orders to deny the authority of central government. When this defensive advantage was combined with a national political consciousness which infected the ranks of the army, it became a powerful force. The Parisian barricades of 1792, 1795, 1830, 1848 and 1871 certainly influenced the path of France's polity. They could not, however, withstand ruthless and concerted military action, as Napoleon showed with his artillery in 1795. The British Army repeated the lessons of Paris in Dublin in 1916. In the following year the refusal of disenchanted soldiers to quell riots in the streets of St Petersburg precipitated the Bolshevik revolution. Behind their barricades, the militant factory and transport workers of the city, maddened by war and hunger, formed the base from which revolution and the soviet structure spread so as to dominate Russia by the end of 1917.

In the Franco-Prussian War of 1870 Paris had merely been besieged and starved — the fighting in the streets was civil strife. In the war of 1914-18 little fighting took place in cities. By this juncture, the spread of houses and buildings over the fields which had begun in Great Britain around 1750 was well under way in Belgium, the Nord, the Ruhr, Silesia and on the fringes of national capitals. The experience of World War I and a recognition of the growing significance of urban occupations to national economies and power led Liddell Hart and

Douhet to formulate the theory of strategic bombing.

During World War II the cost of fighting in cities and the costly damage which any conqueror would have to make good to enjoy the fruits of victory, led to a mutually acceptable limitation of war and the explicit or implicit declaration of open cities in the cases of Paris, Rome and Manila. MacArthur withdrew to Bataan and Corregidor to avoid the destruction of Manila and protract his defence. On MacArthur's return to Luzon, Yamashita declared the city open, though some Japanese troops not under his command indulged in some mayhem and fought in the city. In 1940, the Wehrmacht rapidly took over the dense urbanity of the Netherlands with a combination of paratroopers and tanks capturing the key points in this corridor of conflict.

The unintentional bombing of civilians in London in mid-1940 led Churchill to retaliate with an attack on Berlin. Hitler came back with the blitz, putting Douhet's theory into practice. Whether the motive for this was purely revenge or if the intention was to terrorise the civilian population is unclear. What is clear is that only Goering saw it as a prelude to invasion, which neither Hitler, his admirals, nor generals took seriously. This bombing of cities and the allied response proved a strategic failure and it has been credited with strengthening national resolve among the bombed. Despite this the potency of strategic bombing became a matter of doctrine and was pushed technologically to the extreme of annihilation in Hiroshima and Nagasaki. It provided the teeth for massive retaliation and has now reached the pointless limit of mutual assured destruction, with the principle cities of North America and Europe as the hostages to this uneasy stand-off.

The experience which coloured the attitudes of all major armies to fighting on the ground in cities was the battle of Stalingrad in the autumn of 1942. Hitler gave Paulus and the 6th Army the task of capturing the city. Being of little strategic significance this seems to have been a symbolic gesture to prove the superiority of Germans and national socialism and fulfill the words of an old proverb which predicted that whoever crossed the Volga would conquer Mother Russia. The initial German advance from the edge of the town through the residential sectors to the centre of town was comparatively easy despite the time-buying resistance of Russian storm groups in house to house fighting. Within a week they were penetrating the centre of town and within two more weeks had taken it. The impenetrable defence was met in the factory districts of the inner city where the defence could employ its massive buildings to advantage and the close fighting and dust and smoke negated German air superiority. Here the Russians kept a foothold draining Paulus' supplies and ammunition in the face of coming winter and the advance of seven Soviet armies, totalling one million men. These encircled and trapped the 6th Army.

The Russian sweep westwards taking Kiev, Leningrad, Warsaw and Berlin with overwhelming superiority of numbers, left a quantitative rule embedded in Soviet military theory. This holds that whereas a 3:1 manpower superiority is necessary to take a defended position in the country, a ratio of at least 10:1 is required to ensure victory in a city. The Soviet army's next major experience was in Budapest in 1956, where it took four days of street fighting and the killing of 25,000 Hungarians to suppress a revolt.

The Allied armies saw fighting in Cassino, Brest, Arnhem, Aachen and the Ruhr. Since then, the US Army has had a little practice in Seoul and in Hué during the Tet offensive of 1968. Whether this latter held many lessons for fighting in a modern metropolis is questionable. It is part dense Annamese copy of the city design of imperial China and part 19th century French colonial cantonment. The British Army combined police duties with skirmishing in counter-insurgent operations in Jerusalem, Nicosia, Aden, Belfast and Derry. The French did the same in Algiers. The German army carried memories of the rear-guard defences of 1945 and more than any other was conscious of the likelihood of having to fight a defensive war in their own cities in the future.

Whatever the realities of experience, analysts and the military establishment have concluded that the advantage lies more heavily with the defence in an urban setting than in a rural one. The ratios they foresee may not be identical to the Soviet ones, but they have the same order of magnitude in mind. From this it is concluded that the first law of urban warfare is still that promulgated by Sun Tzu: 'Stay out, go around!'

This view seems to be founded on an impression of tall, densely packed buildings, providing defensive cover and, subsequently, tank traps of rubble. This may still hold good as a characterisation of city centres as built or rebuilt in the old style. It is significant, however, that when Baron Haussmann wished to make central Paris less defensible by the working classes, he employed design criteria similar to those adopted for 20th century suburbia. He drove boulevards into the heart of the city making it accessible to light, fresh air and infantry. The proportion of roadway and open space to buildings was increased to reduce the defensive potential of the traditional barricades. Despite the radical difference between older city landscapes and the new suburbia, the NATO establishment appears convinced of Clausewitz's fourth principle of strategy which proclaimed that the assistance of the theatre of war accrues to the defence. They are sanguine that the vast extension of the built-up area in Europe since 1945 will slow down the rate of advance of any aggressive drive by the armies of the Warsaw Pact. The current doctrine of the US contingent of NATO is given in the US Infantry School's *Combat in Cities Report*:

> The vast increase in the size of Western Europe's urban areas has presented the US Army with a unique situation in the defense of the region. The urban sprawl of the Ruhr Valley, Hamburg area, tied in with the natural defense position of the Taunus Mountains, coupled with the Frankfurt area have provided the defender with large build-up areas (obstacles) the aggressor must pass through to continue his advance.

The spread of residential and commercial occupance of land since 1945 has mostly been in low density suburbs planned around the use of cars with an intense network of good roads, laced together by limited access, divided highways. Houses are detached or in small clusters with an abundance of space about them. They are mostly one or two storied structures of comparatively flimsy construction. There are enormous tracts of land presenting this man-made terrain in some of the strategically critical areas of the world, especially north-west Europe. It can be argued that such suburban terrain is easier to attack than to defend in tactical terms. If such is the case, the strategic implications is that a swath of suburbia would constitute the best line of advance, for a Soviet attack on western Europe for example. This would put paid to the NATO hope that the urbanised landscape would hinder Warsaw Pact forces sufficiently to allow for a negotiated cease-fire before they reached the Atlantic. In order to test that conjecture, it is necessary to demonstrate the potential penetrability of suburbs.

The Evaluation of Urban Terrain

For military purposes the natural and man-made elements of the land-scape are analysed to assist a commander in making tactical decisions to use the land to his advantage and against the enemy. Reduced to fundamentals, for an 'attacker, the selection of a line of advance depends upon evaluating the penetrability of terrain among the options available. Penetrability arises from the balance of mobility and cover afforded by the landscape. The velocity of penetration is conditioned by the advantages the terrain presents for mobility, which fall to the aggressor. The other principal governing factor is the amount of cover and detritus a particular terrain affords, which presents the defensive advantage. Clearly cover can offset mobility and vice-versa. Conceal-ment, the protection against being seen or detected by instruments which a setting provides, serves both defence and attack. The surprise that invisibility affords to the attack can, however, prove crucial. To be conservative we can consider it a neutral factor.

Taking the mobility element of the balance first, it is evident that the ease of winning territory by force of arms is not merely a matter of how direct the routes available to reach an objective from a starting point are and their capacity to carry traffic. It also depends on the ease with which different approaches may be taken if opposition is met on one line of advance. For the same opposing force, this is clearly a simple function of network density. The denser the network, the greater the ease with which any opposition can be by-passed or out-flanked. To take the simplest case of a square lattice network, the additional time taken to traverse a given distance if opposition is met on one route in that lattice and can be circumvented, is equal to one link length. This distance will clearly be less the finer the road grid and the denser the network. From this viewpoint alone then, urban areas with an increasing intensity of road space towards the centre should provide an attractive avenue for attack if speed is of the essence. The relationship of mobility to distance from the urban core should be negative as in Figure 9.1.

It can be taken for granted that any attack spearheaded by tanks, such as might be expected from the Soviet collection of hardware, would stick to the roadway as much as possible. Speeds over fields are less than half those which can be achieved on metalled surfaces, and even worse in wet weather. In this case, from a defensive viewpoint it seems evident that the same force can defend more frontage in rural settings with sparse networks than in the dense network of urban terrain.

From the defender's perspective, urban terrain obviously presents more cover and material for blocking lines of advance. The mass of material in the centre of a town can be piled into enormous barriers and the structures can provide high observation and firing points. The penumbra of 19th century row housing which surrounds most central cores provides a considerable potential for barricades and ambush.

Figure 9.1: Penetrability Related to Road and Building Density

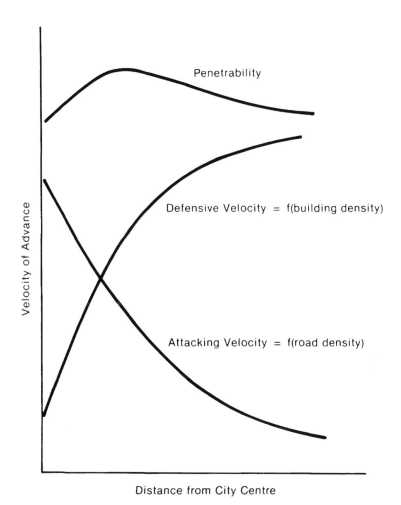

Detached suburban housing, however, is easily outflanked and generates insufficient rubble or vantage to constitute any serious limitation to advancing armies. The basements of standard German houses do provide cover and concealment for defenders but the weapons that could be fired from them safely are limited by the power of their back blast. Although the US Army persists in rating suburbs defensive ground, they do evidently agree that the cover and obstacles afforded in them are limited. This is clear from the tactical training they are providing for their line and staff officers as evidenced in the General

Staff College 1978 manual *Forward Deployment Operations (European Setting) Lesson and Military Operations on Urban Terrain in the Active Defense*. In this, dispersed family housing is presented as having a firing range along streets of 250 yards and a cross fire range of 125 yards. This obviously limits the time defenders have between detecting and responding to advancing forces. On average, roads in this terrain are 15 yards wide, but with big housing set-backs the 'clearance profile' averages 30 yards. This would seem to favour armour. Much of NATO's weaponry is designed for use on open ground at a range of a mile or so and is inappropriate at the hundred yard range characteristic of urban war. The anti-tank TOW cannot be fired in an enclosed space without injuring or killing the operator with its back blast, and its wire-guided missile cannot be brought on target within the average range of exchange in cities. The discarded 3.5 rocket launcher, the bazooka, was more effective in these circumstances. Its short-range replacement, the disposable LAW, will only disable a tank if you hit a track, and has a reputation for injuring its user with its spring while arming it, or for misfiring, leaving the operator with an armed rocket to get rid of.

The Staff College description of suburbia allows that 'obstacles in the streets will be of little value since they can be easily bypassed', 'Armour and infantry approaches are numerous throughout these areas' and 'rubble will have little effect on mobility'. Despite these findings, when it comes to a summary evaluation, the US Army sees suburbia favouring defence by dismounted infantry and as somewhat unfavourable for attack or defence by armoured vehicles. Taking stock of NATO's style and arms, suburbs would, on the contrary, appear to provide increasingly effective cover for an armoured advance. Given the Warsaw Pact's penchant for wheeled vehicles, they will be inclined to stick to the road, and the principal junctions of this network are contained within the urban fabric. There is some evidence that suburban conditions have been taken into account in the design of new Soviet tanks. The concealment provided by low density housing with lots of trees accrues at least as much to the advantage of the attacker as the defence where the range of engagement is so short. In effect, suburbs could conceal an armoured advance as the tall grass and slightly dissected terrain of the veldt hid the advance of Ceteweyo's Zulu impis till they were upon the British at Isandhlwana in 1879.

If we consider only the cover afforded by urban terrain, it is clear that this increases slightly as one passes from rural to suburban surroundings and only reaches a militarily critical level closer in to the centre of a city where 19th and early 20th century buildings or their modern replacements at the same density are encountered. Allowing that the advantage of cover falls to the defence, then the velocity to which defenders can slow down an advance will bear an inverse relationship to the amount of cover. In general it will decrease gently from countryside into the suburbs and then plunge steeply at the radius of the dense inner city, as in Figure 9.1.

Combining the decreasing cover afforded by the urban fabric at greater remove from the middle of town with the decreased mobility afforded by a lighter mesh of roads, the penetrability of terrain may well be construed to increase as the advantages of cover are overcome by those of network density. The military velocity field for invasion would then consist of a caldera-like circular ridge with an inward facing escarpment around the crater of the city core and a dipslope tailing off at the urban-rural fringe. Paths of least resistance through this terrain would be drawn towards the suburban high ground but diverted around the depression of the city centre.

Strategic Implications

If this analysis is legitimate, and the generalisations are representative of German suburbia, then the best line of advance for a Russian army would be along the main east-west corridor of movement between West German cities, sweeping around inner city areas but avoiding open country, employing the advantage of suburbia to the greatest extent possible. This would imply a drive right along the Hellweg, the historical axis of Germany which threads the Börde, the densely occupied ribbon of rich yellow soil that lies wedged against the central uplands at the southern fringe of the European plain. This would engulf Braunschweig, Hannover, Hameln, Paderborn and Dortmund, skirt around the suburban fringe of the Ruhr to capture the Nordrhein knot of the European economy and decant into the intense urbanity of the Rhine delta. An additional advantage of such a strategy would be its nullification of the value of neutron weapons, since much of the advance would be under the cover of suburbanite civilians, not in the rural interstices between towns, where it is intended that the neutron bomb would come into play. The best defence against such an onslaught would be to deploy in the rear of urban cores in such a way as to widen any bifurcation of the advancing columns, extending and dissipating the spearhead of the attack and, thus, increasing the vulnerability of the whole encroachment to fire power. Even then, such a ploy might well be insufficient and suburbia could prove a liability to NATO rather than the defensive boon the US Army presently perceives it to be.

The persistent emphasis in military training on fighting in open country is a matter of the availability of training grounds and, possibly, the shades of Guderian, Rommel and Patton. The romance of a war of manoeuvre on the plains of Germany might well be seen as a modern equivalent of the cult of cavalry and the *arme blanche* which persisted till 1914 in the British Army. This military romanticism, along with a choice of weapons most suited to broad fields of fire does seem like an anachronistic indulgence. Historical experience and precedent cannot teach us in these circumstances. The landscape has been modified in a new fashion over an unprecedented extent. Cities are no longer dots on the map but a large component of the land surface which may provide a theatre for war.

Past masters of the art of war were not faced with this phenomenon. Their general shrewdness, however, can be resorted to for insight into this new situation. For the Russians to go for the cities would be to follow the dictum of Sun Tzu, 'Begin by seizing something which your opponent holds dear; then he will be amenable to your will'. Even if the US Army sees suburbia as defensible ground, the West German army and government must be impressed that 'the decision to attack or defend a city may be tantamount to a decision to destroy it', to quote the American field manual. German suburbia is the object to be defended by deployment far forward, from their viewpoint. A tendency to avoid destructive defence leaves open an avenue of attack that the Soviets have by no means discounted. Soviet military doctrine includes the possibility of 'urban hugging tactics' in order to counter tactical nuclear weapons and employ the accessibility and concealment afforded. The Soviets expect to fight in urban areas on the offence, taking a city every 30 miles or so, and they appreciate the limits which urban angles and ranges impose on the time during which anti-tank weapons can be fired effectively. The present attitude of the NATO military, insofar as it is discernible, is that the use of nuclear weapons is unthinkable and that their task is to stymie any Soviet attack, winning time by an active defence for a political solution to be achieved. In their revealed appreciation of the terrain to hand, they do not appear to be ready for even this limited strategic task.

Urban Revolution

In the Tet offensive in 1968, the Viet Cong aimed to encircle Saigon through the suburbs, not to take the downtown. The aim of this operation was as symbolic as it was military. As part of their political campaign they wished to extend the struggle to the doorsteps of the urban area. The one downtown target was the US embassy, the emblem of American involvement. In the guerrilla campaigns of Latin America there was a decisive shift into the cities in the mid-1960s. A good deal of Castro's success in Cuba, indeed, could be credited to urban support although this has been played down for doctrinal reasons. City slums and indeed fancy suburbs provide better concealment than the mountains. Money, guns and intelligence were easier to come by and targets are denser on the ground. In Uruguay, Guillen pointed out that a guerrilla movement needed the support of 80 per cent of the population and that the centre of power now lay in the great conurbations in which mobile, hit-and-run tactics should supplant the barricades as the mode of revolutionary war. The Uruguayan Tupamaros took up the call but lost mobility in building up a support infrastructure in Montevideo because they failed to win a widespread following. They became an easy target for the military and were mostly put out of action by 1972. In Brazil, Marighela advocated shaking the status quo by urban action in the Rio de Janeiro-São Paulo-Belo Horizonte triangle. He was shot in 1969, but the revolutionary aim of bringing down democratic government in favour of military dictatorship was achieved. In an increasingly urban world, any conflict involving or seeking the interest of the mass of the population is bound to take place in the cities. The future of South Africa is more likely to be settled on the fringes of Johannesburg and Cape Town than in the desert of Namibia.

Not only the commanders of competing imperialisms but also those who seek to overturn from within or preserve the established orders of the world, will find their potential battlegrounds among the streets and houses.

Readings

The relevant military sources on urban warfare are:
US Army Infantry School *Combat in Cities Report* (Fort Benning, Georgia, 1972)
US Army Command and General Staff College, *Forward Deployed Force Operations (European Setting), Lesson 8, Military Operations on Urban Terrain (MOUNT) in the Active Defense* (Fort Leavenworth, Kansas, 1978)

US Army *Field Manual 71-2, The Tank and Mechanized Infantry Task Force* (Washington, DC, 1977) Appendix F

The prospects of suburban war are discussed in:
P. Bracken 'Urban Sprawl and NATO Defence' *Survival* vol. 18, no. 6, (1976), pp. 254-60
B. Bruce-Briggs 'Suburban Warfare' *Military Review* vol. 54, no. 6, (1974), pp. 3-10

10 ZONES OF CONFLICT AND RUMOURS OF WAR

> And as water has no constant form, there are in war no constant conditions
>
> Sun Tzu, (chapter 6, verse 29).

When General Hackett and his fellow NATO generals and advisors contrived a flash point for their fictional World War III, they chose southern Africa. Their appraisal of the world in 1985, as seen from 1977, had the Shah still firmly in command in Iran. In the acknowledgements of the first edition and in subsequent versions of the book, Hackett reiterated the lesson that the only forecast that could be made with any confidence is that nothing will happen exactly as they suggested, although there is a possibility that it might. That Carter and Brzezinski failed to see developments in Iran in 1979, with the best intelligence service in the world and, indeed, good information about what was happening at their disposal, should be sufficient to inject the right attitude of humility into any discussion of the future geography of events from a military standpoint. The events of concern arise from non-recurrent decisions by people, acted out in an arena where chance can tip the balance in a game where there is a strong temptation to play for all or nothing. The inherent uncertainties involved; the labyrinthine logic of human decision, which make it impossible to fathom past choices never mind the future; and the persistence of failure to see trouble coming in international affairs, are sobering antidotes to over-simple generalisation. The possible, substantially different futures we face are numerous and we cannot calculate which will come to pass. We do not really know what happened in the recent or historic past, never mind what may happen in the future, with any precision. Some of the uncertainty we face can be reduced by improving the quantity and quality of information available and the speed with which information is gathered and transformed into intelligence. However, the course of future events must by its nature remain hidden from us in a torrent of prospects which is channelled by certain limiting laws of possibility but remains unbounded in certain directions. Any amount of analysis and theorising will not reduce this torrent to a controllable canal. Preparation for the future then ought to be for a turbulent, churning voyage. To arrange affairs in expectation of a stately progress on smoothly flowing, well-charted waters is to court disaster. In particular to build rigidity into a nation's diplomatic and military stance in the form of an oversimplified map or image of geopolitical relationships is to cut off the paths to many acceptable future states of the world. Inflexibility and a generalised response triggered by an inadequate mental map of the world may easily lead down a helter-skelter path to mutual destruction.

Classes of War

There is a wide variety of violent human conflict. The beginnings of understanding its geography lies in establishing some classes of war which characterise its occurrence. The names given to types of war are laden with emotion and moral implication by association if not intent. 'Holy war' is the most obvious case. 'Revolutionary war' has been applied to the battles of the most reactionary of movements. 'Guerrilla' creates a sense of popular crusade for some and despicable anarchy and cowardice for others. The cold edge of 'nuclear' gives an impression of a technical matter to be decided by the expertise of warring factions of the scientific clergy. 'World war' for total victory was the highest expression of human endeavour against evil for a generation in the British Empire and the USA. That one man's terrorist is another man's hero is evident. It obviously takes as much if not more courage or lack of imagination to venture into an alien setting on your own and plant a bomb as it does to fly over in a plane and drop one. The threat of social annihilation employed in nuclear strategies ought to inspire more terror than individual acts of savagery. If you are going to investigate a vice, then to start by calling some people's sins virtuous is confusing.

In substantial terms the labels applied to wars mix description of geographic extent with political and social involvement and the variety of terrain and tactics employed. Guerrilla war refers to a variety of tactics, urban warfare to terrain, world war to geographic scale and civil war to political limits. The names are often applied in an indiscriminate and inaccurate fashion. Guerrilla is applied to everything from individual acts of political violence to the measured manoeuvering of full blown formations. Some civil wars are fought between social classes drawn from across a nation while others are battles between regional interests. From the Confederate viewpoint the American Civil War was not civil since it was not an effort to gain control of the central government. Frequently, several causes are folded into each other.

To distinguish amongst wars in a fashion which might improve our understanding of their origins and nature, we can picture their position in a three dimensional space allowing for variation in their social, political and geographic characteristics. Along the social scale, the breadth of the social fabric involved can be marked off. At the lowest end of the scale, there have always been politically motivated violent acts by individuals or small groups of the variety commonly referred to as terrorism. Through the largest span of man's history, however, the family has probably been the fundamental fighting unit. This does not just refer to feuding over land or honour. In Belfast and Derry, no matter the formal structure of the Provos or the Ulster Defence League, the basic unit of involvement among the minority directly committed to action is obviously the family. The variety of social categories beyond the family do not necessarily fall into a neatly nested hierarchy. Social classes and ethnic or religious groups may be independent or they may be related so that one social set contains all of a particular ethnic set. It is possible for a battle between economic or affective classes to involve a mixture of ethnic or religious groups, as it did in Russia in the years following 1917. A war between ethnic groups may cut across class lines, as it does in Northern Ireland. On the other hand, ethnic identity or religion may be closely correlated with economic standing as in the battles between indios and ladinos or mestizos in Latin America, or as it was in the Balkan estates of the Ottoman empire. Given this potential complexity, then, a social scale on which to identify combatants ought to be discrete and allow for joint classes. The order on the scale does not necessarily imply an ordered relationship such that each broader category contains some limited subset of prior categories. This becomes even more apparent when we introduce the appropriate spatial limits into the picture. If we use the term 'sectional' interest groups to differentiate a sense of social coherence within a less than national territory, then obviously this sense of belonging identified with one part of a country may cross class, ethnic, religious and even familial lines, as it did in the American War Between the States. Beyond the sectional scope we come to the major identity behind which people have fought each other over the last two hundred years — nationality. Wider than this, we might identify a cultural realm on our social scale, drawing together nationalities with a shared set of traits and mores. This would not, of course, necessarily identify the likely shape of coalitions and alliances. Like nations are frequently involved in competition for resources, territory and power over one another. Alliances frequently cross conventional lines of cultural demarcation. It is one of the ironies of history that William of Orange, whose victory over James II and his catholic followers at the Boyne has come to symbolise the victory of protestantism over the papists, was allied at the time with the Pope against the pretensions of France. The affiliations which various islamic nations have adopted now presents the picture of a rifted cultural domain.

The political axis of this scheme of classification must obviously start at a level less than that of the nation state, where conflict arises from some group seeking to overcome the incumbent authority. To this intrastate level of political agency we must add a trans-national class, to distinguish circumstances where the organisational lines of a power-seeking group transcend national boundaries, as in the case of the PLO. The classic variety of war is the battle of state against state. This may or may not be a clash of two nationalities. Some states are supranational, the United Kingdom and USSR for example. The use of nationality as a propaganda device to secure loyalty is a fairly recent departure, dating essentially from the French Revolution. Nation states have, however, seldom gone to war alone. They have usually sought to outflank or encircle enemies with alliances, and the main danger to our existence now lies here in this category of inter-coalition war, where paranoia reaches fever pitch and sovereignty and responsibility become bent out of shape.

The geographical scale of operations or involvement in a war, our third dimension, rises up from the local to the regional, the inter-regional, the -continental, the intercontinental, to the global. Any one conflict can climb this ladder of geographic scope, embroiling more and more of the world's population and heightening the danger of nuclear conflagration of much of civilisation. The great power game now is to seek to get your way over the matter at issue while containing any actual fighting within the narrowest geographic limits possible to diminish the prospect of escalation. However, the threat of escalation has been employed and lies readily at hand.

The geographic labels — urban and rural — which have been used to describe warfare, do not distinguish a geographic scale but, rather, the variety of terrain encountered. This encounter can take place in a variety of social and political settings and within a wide range of possible spatial limits.

Any analysis of the frequency and distribution of war might well employ a classification which at least distinguished these fundamental elements of social involvement, political scope and geographic scale and would do well to avoid a typology based on ideological name-calling or boasting. The motives for wars would present ground for dispute as a basis for classification but clearly control of population, territory and resources is the underlying objective of them all. A geographic generalisation on the distribution of wars arises quite obviously from the clusters which emerge between the three axes. War arises from some form of political imbalance. Dissent within a nation is overcome by stirring up an external danger to close the parting seams of national coherence. A mismatch between the national territory and resource base and the economic and political energy of a nation can jeopardise peace. Neither Japan or Germany had a lightly peopled expanse in which to seek their versions of manifest destiny in the 1930s and 1940s. On the other hand, a seriously overbounded nation, whose borders encompass too great a range of disparate interests, may seek a common foe to tighten the circle. A nation, such as Russia, which has known conquest and destruction, may be driven by fear of chaos to expand control of its perimeter in the search for a defensible frontier. Political instability can arise because the rising expectations of a people in transition for a traditionally or colonially organised economy have come up against an intransigent elite clinging to their prerogatives. Where the uncertainties in such a setting threaten to disrupt a major portion of the energy supply for the industrialised world, then we have a potentiality for explosion. The temptation to precipitate a dislocation of supply to discomfort a potential foe is nearly as great as the enticement on the part of oil users to secure supplies by force of arms.

In trying to predict where the greatest danger of war lies, it is evident that extrapolation from the events of the recent past is not likely to provide a foolproof divining rod. The events in question are not random occurrences in time or space. To set outbreaks of war in objective categories and examine the circumstances associated with them should aid understanding and inform the judgement of potential peril. Obviously, in such an exercise, frequency has to be weighed by some factor of importance and the number of people killed seems a properly negative expression of the significance of war to mankind. This then raises the problem that the potential death-dealing capacities of the weapons of the USA and USSR have never been employed and that the possible outcome cannot be gauged from any prior observation. The bombs dropped on Nagasaki and Hiroshima were tiny prototypes of present weapons. Everything else pales to triviality beside the prospect of a nuclear exchange. In the last resort other wars only signify inasmuch as they contribute to or diminish the likelihood of this madness. Thus, some wars may indeed have a net human value for all of mankind.

If a localised combat between the interest of NATO and the Warsaw Pact avoids all-out confrontation it may have reduced the horrifying prospect of hundreds of millions of deaths. That is providing its outcome does not leave one side or the other in search of revenge or seeking to redress the balance struck at its close. This cannot be judged at the outset, so the obvious risk of escalation might well suggest a prudent pacification as the safest behaviour.

To seize every outbreak of war as an opportunity for big power competition and to dabble in minor unrelated conflicts to discomfort the opposition is a long-standing ploy of military oligopolistic competition. The economic and technological advances which have reduced the contest to a virtual duopoly, introduce the radical instability of two-sidedness into the game. Although in terms of political institutions a two party arrangement is held to be stable on the basis of the US and British examples, true two-sidedness leaves no room for compromise when the prize is all or nothing. The US polity has in reality more like fifty parties and in Britain the real threat of a third party taking the middle ground, despite first-past-the-post elections, maintains a degree of central tendency in British politics. In military terms, as economic development and political evolution brings more power to Japan, Europe, China, India, and the nations of the Middle East, Africa and Latin America, so a greater diffusion of power may bring a return to the limitations on war which preserved European civilisation from the Treaty of Utrecht to 1914. Competition between two parties seeking to ensure their dominance of the largest extent of the globe necessarily encourages ideological extremity. If the cost of retaining loyalty is very high, then each competitor will wish to maximise the certainty with which they can count on their client states. By offering a political platform like their competitors, each increases the possible geographic extent from which they can attract followers but reduces the certainty with which they can count on them, since the differences become blurred. Taking an extreme political stand reduces the extent of the appeal but secures an assured following from a goodly part of the domain. To promote collective, centralised solutions to the problem of poverty generally attracts those in the most desperate straits, while espousal of *laissez faire* mostly appeals to those whom the state of trade currently favours. This assured following, once committed, is then less costly to maintain with bribes, concessions or force. This tendency to extremes does hold a glimmer of hope inasmuch as it opens up the middle ground for a doctrinally moderate newcomer to step between the antagonists and capture their inner flanks and, possibly, a major share of the attention.

Flash Points

Whatever we might venture to predict in generalised abstraction, the main areas of danger on the globe are the north European plain, the oil fields and routes of southwest Asia and the lowlands of the Amur and Assuri on Eurasia's eastern flank. These are places where big powers feel threatened enough by each other's proximity to vulnerable territory or resources to react violently to a real or imagined danger. There are obviously a great many other zones of friction where actual or impending fighting could build up towards global conflict, but none of these impinges directly enough on the well-being or pride of the major power cores so as to tempt them into facing each other directly in battle with nuclear fire.

As things stand now, the attitudes which inform the overriding nuclear weapon strategy are in a dangerous state of flux. The US conception of an implicit pact of mutual assured destruction (MAD) founded in rough parity of weaponry has been thrown into confused disarray by a perception of Russian superiority. Some powerful informants of decisive opinion believe that Russia's increased numbers of powerful, accurate rockets could take out the majority of Minutemen silos and, thus, the ability of the USA to deliver a second strike. They believe a gap is opening up on the escalator to Armageddon. The spectre of a successful Soviet preemptive strike, surgically eradicating most of America's nuclear capacity with MIRVs, is raised. The Soviets maintain that parity is their objective. Soviet experts scoff at the notion of a neat surgical job and of anything worth having being left if they did such a thing. This realisation and the prospect of dying in the enveloping cloud of radiation are viewed by some as sufficient deterrent. The extreme reaction in US defence circles has been to talk of a policy of launching on warning, threatening to fire everything off on the first remotely sensed evidence of an attack and to postulate a winnable nuclear war. This is seemingly in imitation of a perception of Soviet strategy. It is believed by some that the Soviets were never a party to the MAD suicide pact and that their doctrine was always to be prepared to fight a nuclear war and to seek to come out ahead in this by a combination of deploying more, higher yield, accurate rockets and civil defence. Soviet leaders can, of course, claim with historical justification that they have been the threatened party all along and have only sought to defend themselves from US threatening in the doctrines of massive retaliation and flexible response. They will not exclude the possibility of launching their full might on any warning of a US attack, publically dismiss the prospect of a winnable nuclear war and adhere to a doctrine of deterrence based on the denial of victory to an aggressor. This involves doing everything possible to make sure an attacker cannot hope to survive a nuclear war.

Europe

The geographical occasion for this unquiet turmoil of postures and perceptions of perceptions of perceptions is Europe. The likelihood of either the USSR or the USA attacking the other directly is remote. The US, however, has a commitment to extend its deterrent power to cover its NATO allies in Europe. The threat posed is that if the Warsaw Pact carries out a blitzkrieg attack on Germany the US will be willing to escalate the defence from conventional forces, to tactical nuclear weapons, to theatre nuclear weapons, such as the Pershing, up to the ultimate employment of intercontinental weapons aimed at Soviet missile silos. This is the policy of flexible response. Clearly this threat is empty if US based missiles are vulnerable to a preemptive Soviet strike. The US would then be reduced to defending Europe with its bombers and submarines directed at Soviet cities, which would call down a Soviet attack against American cities. It is deemed by many that as a threat this is not worth a bucket of warm spit, because neither the Russians nor west Europeans believe that a US president would sacrifice millions of American civilians for the sake of keeping Germany and the Low Countries out of Soviet hands.

The window opened up by the perceived superiority of Soviet nuclear forces was met with the stop-gap NATO decision, in December 1979, to place Pershing II and cruise missiles in Germany, capable of reaching all of eastern Europe and some of western Russia. These were designed to offset the SS20s the Russians had stationed to point at Germany. This deployment by NATO is presented as a means of filling the gap between tactical and intercontinental nuclear weapons to assure NATO allies that the US will not take advantage of a firebreak to uncouple itself from any European conflict and avoid escalation to the intercontinental level. No US president has, however, explicitly voiced willingness to risk urban America to keep the Russians out of the Ruhr. Indeed, Reagan's opening ploy in arms control negotiations with the

USSR in November 1981 was to propose the 'zero option' for Europe. This involves giving up the planned deployment of Pershings in return for Russia's scrapping of all of its medium range missiles wherever they are located. A less demanding position would involve Russia withdrawing its SS4s, SS5s and SS20s out of range of western Europe. The last resort would be a trade involving Russian withdrawal of medium range weapons beyond the Urals, from whence they could still strike some western European targets. The Soviet short-range SS22, which does not come into the negotiations, with a range of nearly 600 miles could hit targets in West Germany, Denmark, Norway and Northern Italy from sites in Russia. These missiles can be moved by transporter at 25 mph and so could be moved into East Germany in hours, bringing all of Europe, including Great Britain into range.

In March 1982 Brezhnev responded by announcing a freeze on the deployment of SS20 missiles west of the Urals and the suspension of a programme of replacing obsolete SS4 and SS5 missiles now pointed at western Europe with SS20s. Since the Soviets have already deployed the 300 SS20s they intended to deploy in Europe and can hit European targets from east of the Urals with this missile, this wins a trick in the propaganda game with no sacrifice. Brezhnev's suggestion that the extent of the oceans patrolled by both sides' missile submarines be limited and that the further deployment of cruise missiles be frozen also work to the disadvantage of the USA which is more heavily dependent on its submarine leg and has not got its cruise missiles in operation, which the Soviets have. The threat which went along with these seemingly generous moves was that if new missiles were deployed in western Europe by NATO, Russia would be tempted to put the US in an 'analogous position'. This has been interpreted as meaning that the Soviets would attempt to install missiles in Cuba or Nicaragua.

The combination of uncertainty and the renewed awareness of the dangers of nuclear weapons has made many people in Europe extremely nervous and inclined to opt out of the nuclear game. Opinion polls taken in April 1981 suggest that half of the British population and two thirds of Belgians and Netherlanders are opposed to the deployment of cruise and Pershing weapons in Europe. About a third of all West Germans are against the missiles while 70 per cent of those under twenty were opposed. In Britain, the numbers in favour of unilateral disarmament have risen from less than a quarter to a third between 1980 and 1981. A majority of the French would prefer to stay out of a war between the USA and USSR and 40 per cent would prefer to be neutral rather than allied to the USA. It appears that a majority of

Europeans do not see any great danger of a Russian invasion and in any case do not favour a defence against this involving nuclear weapons.

The longer term plan of the US to maintain some kind of intercontinental parity involves the MX missile and some means of enabling a sufficient number of them to survive a Russian strike to constitute a threat. All of this structure of fear and meance from the US side is predicated upon a belief that the Kremlin's hidden agenda includes the conquest of Europe. Although Marxism played its part in Khruschev and Brezhnev's foreign policy, imparting an ideological tilt in favour of change rather than the status quo, there is no evidence of an underlying master plan of world conquest. There is, indeed, a greater compartmentalisation in the Soviet constitution and political practice, separating military and foreign affairs than is the case in the US executive. Of late there have been indications of a growing concern with predictability and stability in world affairs in Soviet councils. This desire for greater certainty found expression in impatience with the Carter administration's zig-zags in its dealings with the world.

The growing pacifism of western Europe and the steadying of Turkish, Italian, Greek and Spanish societies despite continued turmoil, would seem to offer the US the prospect of retracting its sphere of deterrence. If Yugoslavia and Greece remain calm, the prospect of stability on the western peninsulas of Eurasia is good and the excuses for Soviet interference are limited. The rational thing for the US to do in terms of everyone's chances of survival is to cut a firebreak that prevents the prospect of a trivial spark starting the nuclear holocaust.

Since the Russians are having enough trouble containing their satellite states now, there is no reason, other than doctrinal rhetoric, to believe they would want to extend their imperial territory. The prospect of fighting the Polish army, even with its extremely limited supply of ammunition, seems to have given Russia cause for pause in the summer of 1981 over military intervention aimed against Solidarity. The political actions of the Polish army in December left the USSR with no excuse. The mainspring of Russian fear and aggression in western Europe is deeply rooted in a history of attacks from the west. Most recently, Russians have twice suffered the attentions of an aggressive, expansionist Germany coming at them from the west. West Germany has deported itself in a fashion which should have allayed that fear somewhat. Significant numbers of people in Germany, the Netherlands, Denmark, France and Italy and the UK have expressed a wish to be neutralised or 'Finlandised'. Germany, in particular, has been extremely conciliatory in its *Ostpolitik*. The main danger then lies in the satellites which Russia needs as a loyal buffer. Populist movements in these, such as Solidarity, and the growing unrest in Rumania contesting the party dictatorship and potentially infecting the population of the USSR itself, are seen as endangering tranquility and deserving suppression. At this juncture there is a perceived danger that NATO may seek the opportunity to intervene with a destructive strike in pursuit of the overthrow of communism, despite all the talk of co-existence and detente. Poland is not the extreme case because East Germany lies between it and the west. Should trouble arise in East Germany or Czechoslovakia, the Soviet fear and reaction might be greater. Along the Danube corridor it has the Austrian buffer beyond Hungary.

East Asia

On the Asian edge of its extended Pacific perimeter, the USA has discovered, after much painful experience, that Indochina, a backwater with no great resource base, is strategically irrelevant to her interests. The myth of Chinese expansionism has been defused and the strength of independent power centres in the region is manifest. Although North and South Korea continue to bristle, China seems bent on seeking an accommodation with Taiwan. China and the USA have been cooperating ever more closely since 1972 in matters of intelligence, technology and economics. In July 1981, Secretary Haig announced the lifting of a ban on arms sales to China. Meanwhile China and Russia have been making the first moves towards a possible reconciliation. This would provide a more complete network through which each could pursue its best advantage diplomatically in negotiating with the USA. China's cultural genius has survived for thousands of years by eventually getting its way in treating with barbarians. To expect China to be content with being played as a card in the dealings between the big two is simple-minded. China's tradition is one of pragmatism not ideology. Obviously, if it can reach an accord with the bastion of capitalism for its own benefit, it can as easily cooperate with an ancient enemy, erstwhile patron and recent doctrinal disputant like Russia.

Elsewhere in the region, the coherence of Filipino society should weather any political stability quite well, and Indonesia has displayed a similar capacity over the last several decades. The growing Russian presence in the Pacific may irritate Japan and the USA but this does not seem to have a focus of discontent. Japan's politicians see no real threat from Russia and have no intention of readying themselves to do battle. Two thirds of Japanese citizens are happy with their current modest level of defence spending. Japanese leaders regard their difference with Russia over the Kuril Islands as a local matter and would prefer it if the US did not make a big deal out of it. The dangerous flashpoint of Eurasia's eastern flank is in the valleys of the Amur and Assuri where thousands of Russian and Chinese troops glower across their border with each other. The USSR has a quarter of its army and air force stationed along the border with China. The new railway, the Baikal Amur Mainline, will provide for readier logistic support for this theatre. We have touched on the historic origins of this opposition and it is clear that national pride and not ideological hegemony is at stake here. Inasmuch as the USA has begun to play the China card in its bid to encircle the Russians, it is militarily entangled in this contest already. When China can wield a substantial nuclear weapon then this may be a source of great danger as the USSR deploys SS20s in the region to oppose any Chinese rockets.

The clashes which China had with India were again historically motivated, and not examples of aggressive ideological empire building. It seems to have been the case that India's shrill desire for recognition was the main culprit in sparking combat. The easing off of the Han imperialism which Mao promoted in Tibet, Singkiang and Inner Mongolia, reduces the tensions which might have reached the limit and broken out violently around these border lands of China.

South Asia and North Africa

As long as they do not acquire extensive arsenals of nuclear weapons, nobody is much bothered by what India and Pakistan do to each other. Pakistan, of course, represents a stopping point according to the Truman doctrine definition of spheres of influence, should the Soviets encroach across the Hindu Kush. Afghanistan was, however, placed within the Soviet sphere. Why the Soviets would want Pakistan is unclear, although they have crossed its borders in pursuit of rebellious tribesmen. The way down to the Indus via the Khyber and Khojak passes is hard and dry while routes through Baluchistan are long, over waterless deserts and a very roundabout way of getting into a position to control the Strait of Hormuz. Soviet action in Afghanistan has only elicited bluster from the USA. The main direction of danger to the general peace here is in the spread of dissent to the Turkic moslems within the USSR and the ensuing Slavic panic. The Soviet Union has been encouraged in its involvement in south and west Asia by the adventurous spirit of its Turkic population and their cultural kinship with peoples beyond its borders. The acquiescence of the official islamic clergy gave the USSR the reputation of being compatible with Islam. The Soviet occupation of Afghanistan had support from Soviet moslems. Beneath the official veneer, however, the grain in the Caucasus and the Turkic republics runs to fundamentalism and nationalism, organised around the outlawed Sufi brotherhoods. Events in Iran, the resistance of the mujaheddin in Afghanistan and propaganda from China's Singkiang province directed at stirring up moslems has already caused a Russian anti-islamic reaction, which may well provide the setting for violence.

Given the difficulties Iran presents to native efforts at governance, what ruler in his right mind would want to pick up such an unruly province, even with its oil? Even so, since the ineptitude of the USA has left a vacuum, the communist camp has taken every opportunity to ingratiate itself. Soviet moslems are advising on the suppression of the urban guerrillas and the mujaheddin, and East Germans are helping in attempts to restore the workings of the economy. The war between Iran and Iraq has been left severely alone by the great powers, as it provides an object lesson on the limits which terrain and cost places on conventional fighting in these times. This conflict and the instability of the regimes of both countries are cited as the excuse which would provide the Soviets with a path to the Gulf and control of the premier oil source of western Europe. The depleting resources of the Soviet homeland are held up as a more positive motive than the desire merely to disrupt the economies of NATO. The myth of the compulsive Russian desire for a warm water port is trotted out as an additional motive.

What drew the USSR into the Middle East in the 1960s was a matter of strategy. To face and check the challenge of US missile bearing Polaris submarines the Soviets extended their naval coverage, including the commissioning of a Mediterranean squadron. This revitalised a longstanding Russian desire for right of passage through the Dardenelles. The Soviets' decision not to build aircraft carriers led them to seek air bases ashore from which to cover their ships, first in Egypt, then Syria and then Libya. In an effort to secure its footing, the USSR courted local military and political support and negotiated treaties. Although there was an ideological impulse from some quarters in the Kremlin to engineer communist take-overs of government, this was suppressed where it interfered with the strategic objective. If there was little risk of

American opposition and the opportunity arose, as in Aden, the impulse was given rein. Given the retreat of the British imperial presence and limited US interest in the region in the 1950s and 1960s, by proferring generous aid and using a light touch politically, the USSR diffused its influence widely. The Arab need for arms and support in their conflict with Israel provided the occasion of penetration and the air and sea bases from which to serve its strategic purposes. US support of Israel more clearly identified the USSR as the Arab's benefactor.

As the Soviet presence became more imperial, so the welcome cooled. In 1972 Sadat threw Soviet advisers out of Egypt and rescinded the friendship treaty between the two countries. Syria refused such a pact, leaned heavily on its local communists from time to time and went its own way on Lebanon. Iraq denied the Soviets a naval base at the head of the Gulf. The quarrels and the shifting positions of Syria and Iraq, of Egypt, Libya and the Sudan and of the PLO, Jordan and Lebanon, have made it difficult for the USSR to maintain its position throughout the area. It cannot be on both sides of a dispute and retain the loyalty of all parties for long. At this scale the inverse of divide and rule seems to apply. You cannot retain the allegiance of a deeply divided people. The emergence of Saudi Arabia with its oil wealth as an arbiter of affairs reduced Soviet influence. With the assumption of power by the Ayatollah Khomeini, islamic fundamentalism poses an appealing counter to Marxist-Leninism. The moslem socialism of Libya and its capricious leader is not wholly compatible with Russian interests. The megalomanic activity of Col. Qadaffi is obviously beyond the Kremlin's control and his support of insurgency does constitute an embarrassment in some cases. It has been suggested that Russian training and weapons support a network of disruption and terrorism, placing the gun on the Libyan table, available there for any group which will irritate NATO. On the other hand there is some evidence of the involvement of US trained personnel in Libyan terrorist activities. The stratagem of fostering terror was used by the SOE and OSS in World War II and, subsequently, the CIA, to embarrass their nation's enemies. It proved to be an unreliable and uncontrollable weapon, liable to backfire. Col. Qadaffi's venture into Chad seems to have been powered by a desire for sheer territorial aggrandisement and the opportunity presented by the chaotic strife within that country. It was hardly motivated by a desire to increase Libya's resource base. To construe it as a domino-like extension of the communist empire through the agency of Qadaffi, directed across the Sahara at West Africa seems somewhat extravagant, especially now that Libya has been forced to withdraw from this expensive venture carried on over 1,000 mile supply lines. The dissent generated by the cost and conscription for the war in Chad is eroding the Colonel's authority. Another of his enterprises involves support of the Polisario guerrillas in the western Sahara, along with Algeria. The Algerians would like a compromise settlement in which Morocco retains the northern part of the former Spanish colony and its phosphate deposits, while the Polisarios retain the south and form a federation with Mauritania. Although the Moroccans claim that the Polisarios have Russian equipment, there is no evidence of Soviet interference. Algeria's conciliatory attitude has opened a growing rift in relations with Libya. They now have a territorial dispute and there are accusations that Qadaffi is fomenting trouble among Algeria's nomadic Taureg population. All of this amounts to local conflict in which neither great power stands to gain much by interference.

The same strategic consideration which brought the Soviets into the Mediterranean and the Arab world has now pulled it further east and south to curtail the degree of freedom enjoyed by US submarines in the Indian Ocean. The Horn of Africa is a commanding location from which to supply and direct operations in this ocean, in the Red Sea and the Persian Gulf. In the early 1970s the USSR acquired the use of Berbera from Somalia, where President Said Barré needed arms for his battle with US supported Ethiopia for Ogaden, an Ethiopian province peopled by Somalis. When Mengistu seized power in Ethiopia in 1974 he dismissed the US connection, tainted by association with Haile Selassie, and sought Russian help to repel the Somalis from Ogaden and crush the rebellion among the moslem peoples of Eritrea, Ethiopia's front on the Red Sea. Caught between warring clients, the Russians and their Cuban allies provided arms and troops to Mengistu. The Somalis turned to the USA and Egypt for help and offered the evacuated Russian base of Berbera for the use of the Rapid Deployment Force. The US has been leery of Barré's deviousness and of being associated too closely with Somali aggression. Russian support of traditionally christian Ethiopians against moslem Eritreans cost it support in Syria and Iraq. Russian and Cuban intrigue against Mengistu has lost them much influence as he has strengthened his hand and scope of independence. Ethiopia, Libya and South Yemen signed a mutual defence treaty in August 1981. At the same time, Ethiopia and Sudan are informally allied: Sudan offering help against Ethiopian rebels based in the southern Sudan and a friendly nation in the ring of US clients which encircles Ethiopia in return for Ethiopian restraining influence on Libyan territorial ambitions which were directed at Sudan from Chad. This all suggests that as the USSR extends its perimeter of involvement, the complexity of conflicting and mutual interests within that radius dissipates the potency of its presence.

The suspicion of a thirst for oil as the driving force in Soviet doings in the Middle East has been fed by the US State Department. Intelligence reports which are broadcast must always be greeted with doubt as to their intent. Published CIA estimates of future Russian need for Gulf oil must be suspected of exaggeration. Among other things, to become dependent on a remote resource would violate Lenin's institution of autarchy. There is no evidence of Soviet plans to import large quantities of oil from the Gulf. Her European allies have been encouraged to buy more from the Middle East, especially since the USSR demands dollars for its oil. The great bulk of east European needs, however, are still provided by the USSR. Any temptation to disrupt the oil supplies of Japan, western Europe and the USA carries the risk of a direct military clash with the USA.

The best evidence of the Soviet stance in the region lies in its behaviour in the Arab-Israeli affair. Although it seems that Gromyko stirred the Six Day War in 1967 in order to consolidate the Soviet position with the Arabs, since that time the prospect of coming face to face with the USA has led, by and large, to the Russians restraining conflict. In 1973, the Yom Kippur War finished with both the USA and USSR effectively imposing a ceasefire. To reduce the prospects of an eruption in the region the Kremlin has worked for an Arab-Israeli settlement in which it is formally involved, thus confirming the legitimacy of its presence in the region. In the meantime it incurs the wrath of its radical clients over this – Syria, Iraq, Libya and the PLO. However, its continued relationship with them is its entrée to a powerful position in any negotiations.

The focus of Soviet political involvement appears to have followed its strategic needs southwards. Policy makers in the USA have chosen to interpret this in terms of territorial ambition rather than of strategic deployment and have built up a picture of Soviet designs on the oilfields and seaways of the Persian Gulf. Three direct paths for Soviet assault on the Gulf have been identified. The first is through Baluchistan and along the Iranian coast to a position commanding the Strait of Hormuz. The second is across the Persian Plateau and the Zagros to the Shatt el Arab at the head of the Gulf. Operation Blue, the attempt to rescue US hostages from Teheran, demonstrated the inhospitability of this terrain. The third is from Armenia to the headwaters of the Tigris and thence to the head of the Gulf through Iraq. None of these are easy and none would be without local opposition. The indirect approaches which the US suspects, involve the fostering of revolt in Saudi Arabia, the Trucial states or Syria. The prospect of continuing Arab-Israeli conflict in the region is a perennial source of instability. Whether Mubarak in Egypt can withstand Islamic revivalism and continue to diminish Russian influence in the region remains to be seen. It was this array of fears and doubts which led the US to try to solve its logistic problems in this region far from its shores with the Rapid Deployment Force to act as a trip wire to discourage Soviet advances. The geographic breadth of this theatre of conflict spreads a long way from the oil fields of the Persian Gulf. The Horn of Africa comes into play as a strategic key for stationing forces to command the waters and shores of the Arabian and Red Seas. This extends the scope of diplomatic and potential military action involved to Ethiopia, Somalia and Kenya. Kenya is directly involved in terms of making Mombasa's port facilities and Nairobi airstrips available to the RDF.

The global significance of local conflicts diminishes with distance from the current pivot of history, the oil fields of the Persian Gulf. There is a good case to be made for the meeting place of Eurasia and Africa in this vicinity as the true and lasting geographical axle upon which history turns. Under present conditions the circumference of danger extends from Baluchistan around Afghanistan along the Elburz to the Caucasus, includes Turkey, the Levant, sweeping in Libya, the valley of the Nile and the Horn of Africa. Mackinder's pivot was the region which the horsemen of the steppes swept out of under pressure of limited resources and internecine competition. The prize they most keenly sought lay in the region we have just delineated, now divided among the fragments of Islam, created by these riders from the plains. The source of instability now lies in the confusion of Islam. People who have been sheltered from changes in the west by the conservative, legalistic code of Islam now have to face accommodation with 400 years of technical, social and political evolution all at once. One possible reaction is to deny the morality of what has come to pass and retreat into the past as the Wahabis did and Khomeini would do. At the other extreme is a radical embracing of change in the fashion of Ataturk. Shah Reza Palavi attempted this but did not succeed. Whatever the ploys of government, the people will be caught in the quandary of going both ways at once, and social schizophrenia can lead to savage action and reaction. This imbalance, plus the presence of oil and Israel, turns this into the percussion cap of the world. Both the USA and USSR have contained conflict to the local level here so far. The USA faces the particular strain of meshing its grand strategic objectives with the needs of internal politics and the pro-Israeli faction. Israeli opportunism does not make the task easier.

Southern Africa

In Africa south of the Sahara, Boerdom faces the aspirations of blacks in its thrall and of the unsure beginnings of nationhood in the 'frontline' states. The fires created by this friction will continue to smoulder and flare. The military superiority of South Africa, however, is likely to forestall any effort to drive the white tribe back from the Limpopo for some time to come. Advice and arms from the USSR, Cuba or China are not likely to tilt the balance the other way. The incumbent industrial power is at a great advantage in logistic terms over any technologically comparable power at a great remove or any local, rurally-based opposition. The internal opponents of neo-apartheid are subject to severe repression and are divided, with the superiority of the Zulus working in favour of continued Nationalist supremacy at present.

The consensus of expert opinion on Soviet geostrategy sees its attention focused primarily on Europe. No underlying blueprint for world hegemony can be discerned in the actions of the Soviet government, only a response to seemingly advantageous and low risk chances to expand its field of influence as they arise. The colonial evacuation of Africa left many openings for the insinuation of Russian power on a continent where the interest of the USA was occasional or confused. There are, however, other contenders for allegiance, or at least a respectful demeanour, from emergent African nations. Along with South Africa, France and England still play roles in Africa, as do China and Cuba, and Libya's and Egypt's activities extend beyond the Sahara.

The competition between Russia and China which sprang from their parting in 1959, is in many ways the most active source of turmoil. Cuba plays an independent game in some cases but in others proxies for Russia. In Angola, the wholly black movements with the wider popular following, the FNLA and Savimbi's UNITA, had the more remote support of Zaire, the USA, South Africa and China at various times. Neto's MPLA, with its cadre of urban mulattos, gained the support of the radicals of the Portuguese army and the Soviets. A large force of Cubans were brought in which enabled Neto to control Luanda and most of the country, though Savimbi holds sway in the south. Recent South African incursions from Namibia into Angola, aimed at showing their superiority over Russian supported SWAPO guerrillas before any UN sponsored elections in Namibia, was clear evidence of the absence of territorial integrity. In the convolutions which produced Zimbabwe, Mugabe triumphed leading the Shona based ZANLA, who depended largely on Chinese aid and arms. The rival Ndebele based ZIPRA, obtained Soviet arms and Cuban training in Zambia. Since the negotiated settlement, the guerrilla forces continued their rivalry as two components of the new Zimbabwean army, breaking into fire fights at times. In September 1981 Mugabe caused the Ndebele leader, Nkomo, great unease when he brought in Korean advisers and arms to strengthen the Shona brigades of the army. Continuing tribal war does not provide a stable basis for a polity. The blowing up of Mugabe's political headquarters was charged against the white tribe.

Although they succoured the classical form of the Boer commandos, the grasslands of the South African veld do not offer much comfort for modern guerrillas faced with highly mechanised and air-supported regular forces. The number and expertise of South African white soldiers and their spy network make either rural or urban guerrilla warfare a tough proposition. The total mobilisable force of white South Africa is 404,500. This is twice the army that the seven black nations to the north could field in concert, and far better equipped. The likelihood of success for regular or guerrilla bids for power from outside South Africa is not very great at present. The economic ties of the front-line states to South Africa in terms of transport, trade and emigrants' remittances, hold them hostage to the prosperity of the status quo. There are accusations that South Africa is enhancing these ties by encouraging guerrillas in Angola and Mozambique to sabotage the railways from Zambia and Zimbabwe to Benguela and Beira. Quiescence, born of economic dependence, is also found among the black and coloured beneficiaries of South African growth. Those who prosper within the structure of apartheid, the urban workers, are using the economic weapons of unionism and strikes to gain more power and a greater share of the nation's wealth. Their interests are now distinct from those who remain in rural poverty in the homelands and their tribal elites. Between these, long established suspicions further divide the black power base. Other tribes are especially apprehensive of the growing power of Chief Buthelezi's Zulu Inkatha organisation, which has flourished inside the framework of apartheid.

The combination of strategic mineral resources, the Cape route and racial emotions incline American and European politicians to lean in favour of white South Africa. There are countervailing liberal sentiments which seek the dismantling of white supremacy but this is offset by a temperamental inclination not to upset the status quo entirely and work for gradual adjustment. This leaves an opening for the intrusion of the Soviet Union with more radical campaign slogans and sustains a distinct possibility of direct conflict between the USA and the USSR over South Africa.

Caribbean Basin

In the Americas the greatest current threat to peace is in El Salvador. The USA is in danger of becoming too deeply entangled to step aside if the government begins to topple. The temptation to the Soviets is to encourage Cuban involvement and draw Nicaragua, Honduras and Guatemala into the mêlée. Domino theorists can see a chain reaction leading right to Mexico and the Rio Grande. In terms of a major conflagration, if the USA is logistically disadvantaged in the Gulf, then the USSR, faced with the most formidable navy in the world in its home waters, would be insane to venture too deeply into the Caribbean. If war in this region hotted up, obviously the US could put the Cuban conduit out of operation in short order and control the lines of supply easily.

The only force which could offer a direct threat to the defences of the USA in the Americas is the USSR. Some of its 340 submarines could come within missile striking range through Caribbean waters where the islands offer some cover from surveillance. It is clear that any more conventional attack involving the landing of troops is out of the question even from a base in Cuba. In global terms the Caribbean no longer plays the key role it did in the last century and the Panama Canal is no longer of any real military significance. Its potential importance for commerce has dwindled as the pattern of world trade has bypassed it. Most US trade with Europe goes through Atlantic or Gulf ports. Asia and Europe trade around the Cape and through Suez. US traffic with Asia mostly crosses the Pacific coast. The growth of population of the west coast states above the threshold sufficient to sustain a manufacturing sector of its own has diminished the potential for trade from the manufacturing belt of the northeast and megolopolis to California, Oregon and Washington. The military function, allowing the US Navy to command two oceans with one fleet, has not signified for a long time as the canal is two small for anything but destroyers. What does continue to interest the US in the Caribbean and its rim is the magnitude of its trade with the region, including as it does a quarter of US oil supplies and two thirds of its bauxite.

The weakness of the USA to its south lies in the endemic poverty, social inequality and political instability to be found in some of its client countries. The USA intervened in 1948 to put down a socialist take-over bid in Costa Rica, and similarly in Guatemala in 1954. It has often supported somewhat oppressive incumbents in an effort to maintain a state of rest. Having acquiesced in Castro's overthrow of Batista, a clumsy effort at reinstating the 'liberal' exiles in 1961 by the USA left the way open for the Soviets to establish rapport with Castro and a presence in Cuba. Rebuffed in its efforts to export its variety of revolution to the mainland, Cuba turned its attention to Africa. Despite Castro's lack of success in Central and South America, US influence has also waned. In Nicaragua the ruling Somoza family, which attained power after the withdrawal of the US Marines in 1936, was ousted by the Sandinistas in 1979, signalling a distinct turn to the left in Central American politics. Two years later in El Salvador, the junta, led by Duarte, came under open attack from leftist guerrillas. The insurgent forces are reputedly provided with arms and aid from Cuba via Nicaragua while the forces of the junta have US arms, helicopters and advisers.

There are in this part of the world larger nations such as Mexico, Venezuela and Jamaica which have displayed the capacity to transfer power between political opponents on a constitutional basis, even if they suffer spasmodic political violence. This creates a matrix of stability which may contain any conflagration and keep US intervention at bay. This in its turn will reduce the potential for direct Cuban and Soviet meddling. The Reagan administration is now divided as to the best comportment to adopt south of the Rio Grande. Weinberger and Defense are against getting entangled and Haig at State is for stepping up support of threatened regimes. It is unfortunate that the choice may not result solely from a rational calculation of the best interests of the USA but may be conditioned by the exigencies of not only domestic party politics but White House kitchen politics. The dispute between Argentina and the UK over the Falklands complicated matters in April 1982. The conciliatory attitude of the US administration to Argentina was interpreted as seeking to maintain the support it had cultivated for its anti-communist campaign in Central America. There were rumours that plans existed for an Argentinian contingent to participate with the US in kicking over the Sandinistas in Nicaragua. This was not a pleasing prospect to Mexico or Venezuela, never mind Nicaragua. The Soviets leaped to the support of the anti-communist, military dictatorship and were received with open arms. Even though the Argentinians had used the excuse that they needed the Falklands to keep an eye on Russian submarines, they had also happily met Soviet grain deficits when Carter put an embargo on US grain exports to the USSR in protest over their actions in Afghanistan. It is clear that the peculiar passions of nationalism and self-interest dominate international affairs and that a stone dropped in the South Atlantic can send ripples into the Caribbean, up the Potomac, and even into the corridors of the Kremlin.

Conclusion

One prospect which appears on the horizon now and might bode well for man's future, is that western Europe will choose not to play in the big power game and neutralise itself. Whether or not it was then overrun by communist hordes, there would be little reason for its nuclear annihilation or for the USA and Russia to burn each other over it. In Germany, Denmark, the Low Countries and the UK sentiment in favour of such a choice is waxing. Chancellor Schmidt is aware that large numbers of Social Democrats feel like this. Anthony Wedgwood Benn narrowly missed gaining the leadership of the Labour Party on such a platform and the party itself adopted a unilateral disarmament policy.

On the other side of the Iron Curtain, Russia cannot be firmly convinced of the willingness of Polish, German, Czechoslovak, Rumanian and Hungarian soldiers to fight to conquer western Europe. The USSR may well have some doubts about its Turkic and Slav components' desire to be killed for Marxist-Leninism.

America's accommodation with China and the calm attitude of Japan may diminish frantic US efforts to be the sole and commanding presence in the Pacific basin, rushing into wars it could not hope to win without escalating towards doomsday.

As to the oil fields of southwest Asia, these will pass from hand to hand locally as traditionally muslim people seek harmony with changes in the world which cannot be abolished. Whether conservative or radical, economies and societies in the region are addicted to oil revenues and will be strongly inclined to deliver as long as they can. The fears of direct Russian interference over what seem on a wall-map like short overland lines, should be allayed by the fact that on the ground these are dry, difficult and daunting. To admit that western economies are now heavily dependent on Gulf oil is not to deny that there are substitutes. Western society is not so fundamentally dependent on oil as to be willing to risk its own obliteration to secure supplies of one, substitutable input. Disruptions of supply of the 1970s did not bring chaos and inspired an economic strategy of greater flexibility.

Any accounting of the costs and gains of nuclear war must end up in deepest deficit. In retrospect, the total victory of 1945 appears as no great bargain. If the UK fought to retain its empire and pax britannica, it clearly failed. If the US and UK fought to contain German control over the heartland, they did so at the cost of handing it over to Russia. If the US fought to retain command of the Pacific and break up the East Asian Co-prosperity Zone, the prosperity of Far Eastern economies is now inextricably tied in with that of Japan. American influence in Asia has been won by belated recognition that it was fighting on the wrong side and by reaching an amicable relationship with the Chinese government. If the war was fought to retain Royal Navy and United States Navy command of the seas, it presaged the demise of the RN and the rise of Russian competition. The cost in people's lives, things destroyed, war material and economic and social dislocation was enormous. National socialism was so patently evil that any optimism about mankind's nature would lead one to believe it carried the seeds of its own destruction. The prospect of lasting German domination was built on a Wehrmacht prepared for conquest but not to hold an empire. The danger of invasion of Britain, which stiffened British resolve and drew the US into the war with Germany, was never real. The dragon's teeth of terrorism sown between 1939 and 1945 have yielded a continuing drain on the energies of the victors, now turned upon each other. A judgement on the value of a war after the event does, of course, require the elaboration of a counterfactual alternative. You can only evaluate the decision to do something in terms of what would have happened if it was not done. The costs and benefits are then gauged as the differences between the two trains of events. Historians may deny the legitimacy of 'what ifs?', but such questions provide the only clues to rationality which statesmen can seek. The general answer from such retrospection seems to be that there can be no victory worth having in modern war. Winning is a distinctly pyrrhic achievement. The cost of winning and of refurbishing the defeated turns the spoils of war into a drain on the exchequer.

Perhaps the question over whether to fight or not ought to be phrased in terms of what prize is worth the cost of destroying civilisation, which stands as the all too possible *ultima ratio regis*. On what strategic issue would a chairman or president be behaving rationally to start a nuclear exchange? To prevent German reunification? To keep the Red Army out of Poland? To ensure oil supplies from the Gulf? To prevent a Russian invasion across the Amur? Many of the fears of action on the ground which might unleash madness are greatly exaggerated from ignorance of the protection which distance and terrain still offers in keeping our aggressive drives apart. If we do not let this geographic knowledge calm our fears and allow that the likely outcomes are not worth the risk of destruction, then we will have crowned irrationality and set it to reign with the most unholy terror.

In this book we have attempted to bring a distinctly geographic view to the study of war. Such an analysis can be for the purpose of explanation or illumination of past events, for the design of modes of fighting or for the attempted predication of things to come. As geographers, we have tried to understand the impact of the earth's size and differentiation and man's occupance of a variety of habitats on violent conflicts between groups of people. If this viewpoint adds to the evidence of its ultimate futility, then we will have served a useful purpose.

Readings

A foundational description and analysis of contemporary warfare is provided by:
M. Carver *War Since 1945* (Weidenfeld and Nicolson, London, 1980)

The possible course of a war between NATO and the Warsaw Pact is dramatised in:
J. Hackett et. al. *The Third World War* (Macmillan, London, 1979)

The main news sources drawn upon for this chapter were:
The Economist, Time and the TV networks.

Soviet attitudes were reviewed extensively in the proceedings of the twentieth annual conference of the International Institute for Strategic Studies held in September 1978. These were published under the title 'Prospects of Soviet Power in the 1980s' in the *Adelphi Papers* numbers 151 and 152.
Number 151 included:
R. Legvold 'The Concept of Power and Security in Soviet History' pp. 1-12
A. Dallin 'The United States in the Soviet Perspective' pp. 13-21
J. Laloy 'Western Europe in the Soviet Perspective' pp. 22-9
L. Labedz 'Ideology and Soviet Foreign Policy' pp. 37-45

Number 152 included:
P. Windsor 'The Soviet Union in the International System of the 1980s' pp. 2-10
W.E. Griffith 'Soviet Power and Policies in the Third World: The Case of Africa' pp. 39-46
G. Golan 'Soviet Power and Policies in the Third World: The Middle East' pp. 47-54

The 'Crisis in the Caribbean' was featured in the October 1981 (vol. LIII, no. 3)

issue of *Geographical Magazine* with articles:
R.L. Woodward 'Trouble in Uncle Sam's Backyard' pp. 838-43
S.R. Elliot and I. Lee 'Where the Power is Wielded' pp. 843-6

INDEX

Printed and bound by CPI Group (UK) Ltd, Croydon, CR0 4YY

28/10/2024

01780006-0001